Environmental Footprints and Eco-design of Products and Processes

Series Editor

Subramanian Senthilkannan Muthu, Green Story Inc.
Toronto, Canada

Indexed by Scopus

This series aims to broadly cover all the aspects related to environmental assessment of products, development of environmental and ecological indicators and eco-design of various products and processes. Below are the areas fall under the aims and scope of this series, but not limited to: Environmental Life Cycle Assessment; Social Life Cycle Assessment; Organizational and Product Carbon Footprints; Ecological, Energy and Water Footprints; Life cycle costing; Environmental and sustainable indicators; Environmental impact assessment methods and tools; Eco-design (sustainable design) aspects and tools; Biodegradation studies; Recycling; Solid waste management; Environmental and social audits; Green Purchasing and tools; Product environmental footprints; Environmental management standards and regulations; Eco-labels; Green Claims and green washing; Assessment of sustainability aspects.

Subramanian Senthilkannan Muthu
Editor

Assessment of Social Sustainability Management in Various Sectors

Springer

Editor
Subramanian Senthilkannan Muthu
Chief Sustainability Officer
Green Story Inc, Canada

Tung Chung, Hong Kong

ISSN 2345-7651 ISSN 2345-766X (electronic)
Environmental Footprints and Eco-design of Products and Processes
ISBN 978-3-031-81028-2 ISBN 978-3-031-81029-9 (eBook)
https://doi.org/10.1007/978-3-031-81029-9

This Springer imprint is published by the registered company Springer Nature Switzerland AG
The registered company address is: Gewerbestrasse 11, 6330 Cham, Switzerland

This book is dedicated to:
The lotus feet of my beloved Lord
Pazhaniandavar
My beloved late Father
My beloved Mother
My beloved Wife Karpagam and Daughters—
Anu and Karthika
My beloved Brother
Last but not least
To everyone working in various industrial
spaces to make the world "SUSTAINABLE"

Contents

Assessment of Social Sustainability Management in Various Industrial Sectors

Subramanian Senthilkannan Muthu

Contents

Social sustainability is a crucial aspect of sustainable development that focuses on identifying and managing the impacts of businesses and organizations on people and society. It aims to ensure that current and future generations can enjoy equitable access to social resources, opportunities, and quality of life. The concept encompasses several key principles, including improving overall well-being and living standards, ensuring fair distribution of resources and opportunities, recognizing and valuing diversity, fostering social cohesion, and promoting inclusive governance. In the context of business and industry, social sustainability involves practices that positively impact employees, workers in the value chain, customers, local communities, and society at large. These practices can include providing safe and fair working conditions, ensuring respect for human rights and labor rights, promoting diversity and inclusion in the workplace, investing in employee development and well-being, engaging with and supporting local communities, and developing products and services that address social needs. By integrating social sustainability into their operations, businesses can not only contribute to societal well-being but also benefit from improved reputation, increased employee engagement, better risk management, and enhanced long-term viability.

The second chapter explains the environmental, social, and governance sustainability management and financial outcomes, citing evidence from US tech firms. The role of environmental, social, and governance (ESG) management information has evolved over the years. Some have argued that companies should not allocate resources to socially responsible activities. The previous focus on financial targets

S. S. Muthu (✉)
Green Story Inc, Toronto, ON, Canada

S. S. Muthu (ed.), *Assessment of Social Sustainability Management in Various Sectors*, Environmental Footprints and Eco-design of Products and Processes,
https://doi.org/10.1007/978-3-031-81029-9_1

led to the view that ESG investing was a cost to shareholder value. However, companies have recently argued that they should consider environmental and social responsibilities for market survival. The European Union (EU) has initiated a series of reforms in sustainable finance, transforming how financial and nonfinancial companies operate. ESG practices are a long-term strategy to achieve positive results in company performance by securing social legitimacy. However, there is still a debate in the literature about whether a company's efforts to advance ESG factors improve its performance or reputation. The authors conducted a rigorous panel data analysis to investigate the relationship between ESG scores and the financial performance of US tech companies from 2017 to 2022. The authors conducted comprehensive tests for various statistical issues such as multicollinearity, endogeneity, cross-sectional dependence, slope heterogeneity, cross-sectional time effects, autocorrelation, and heteroskedasticity in the error term to ensure the reliability of our findings. Authors estimated the elasticity coefficients using the panel-corrected standard errors (PCSE) estimator developed by Beck and Katz (1995). The analysis revealed that environmental sustainability management positively impacts financial performance, while social and governance sustainability management has a negative impact. This finding suggests that a company's investments in labor, human rights, community, and product responsibility do not have the expected positive impact on financial performance. The study concludes with a broad discussion of the findings, emphasizing the practical implications for managers, shareholders, and investors.

The third chapter revolves around the social sustainability management practice in the textile industry. Social sustainability management involves practices to manage the business impact on people. The objective of this paper is to identify social sustainability management practices in the textile industry. The method used is a systematic literature review from the academic database Scopus, and the selected articles were screened using the Preferred Reporting Items for Systematic reviews and Meta-Analyses (PRISMA) approach. In analyzing the reviewed literature, the authors referred to social sustainability management practices commonly discussed by prior researchers, including safety working, labor conditions and human rights, society, and product responsibility. Additionally, the authors attempted to identify the impacts generated by social sustainability management practices. This study expands the body of knowledge on social sustainability management by systematically synthesizing various social sustainability practices as discussed in the literature.

The fourth chapter explains how the sufficiency economy philosophy can foster agriculture enterprise for well-being with a case study from Thailand. Globalization has spurred rapid transformations across various aspects of human life, encompassing tangible and intangible dimensions. The corporate sector significantly impacts a company's operations, including consumption and production. Furthermore, business activities have implications for various stakeholders, necessitating an intricate balance between economics, society, and the environment. As such, exploring alternative economic approaches to support corporate sustainability is necessary. This study applied the sufficiency economy philosophy (SEP), introduced by Thailand's King Rama IX, in decision-making processes and its integration within company operations. Grounded in a middle-way approach, SEP can propel businesses toward sustainable development. This chapter evaluates the critical characteristics of agricultural enterprises incorporating SEP into management practices through a three-case study approach. The authors link the results from SEP-integrated enterprises to an improvement in the well-being of stakeholders, particularly employees, and the enhancement of self-sufficiency in areas such as energy consumption and food production. The authors conclude by emphasizing that SEP can guide the business sector toward aligning with sustainable development goals while becoming inclusive firms.

The fifth chapter is about the current dilemma of the social healthcare system in Chile, providing an analysis of the current situation and discussion of the pending reforms, from a social sustainability management viewpoint. The Chilean healthcare system stands at a crossroads, grappling with significant disparities and inequities between its public and private sectors. This chapter describes the dire state of the Chilean healthcare model at present and addresses the current dilemmas faced by the population that stands looking at Chile's healthcare system at a crossroad. This work highlights the inequalities in access, inconsistent quality of care, and financial sustainability issues that plague the public sector while the situation of Chile's private healthcare system is threatened by growing governmental policy pressures and looming uncertainties. By examining the structural challenges and presenting real-life examples, this chapter underscores the urgent need for comprehensive reforms. Thus, through the lens of social sustainability management, the discussion explores how the Chilean healthcare system can evolve to deliver equitable, high-quality services while maintaining long-term viability. The chapter outlines key principles of social sustainability, drawing on successful models from other countries, and provides an overview of the pending reforms proposed to address these systemic issues. The aim is to present a balanced analysis of the potential impacts of these reforms on the quality and sustainability of healthcare in Chile, offering insights into the perspectives of various stakeholders involved. Ultimately, this chapter seeks to contribute to the discourse on creating a more just and sustainable healthcare system in Chile, one that is able to provide quality services at affordable prices in a timely manner for Chile's population, ensuring that all citizens have access to the care they need and deserve.

The sixth chapter dwells into the impact of sustainable technologies on business strategy and competitiveness. This investigation explores the dynamic relationship between sustainable technological innovation (STI) and business competitiveness, a topic that has seen substantial scholarly interest over recent decades. The study highlights how STI significantly influences various business practices, including environmental management, sustainability, transparency, and stakeholder communication. The role of technology as a pivotal catalyst in fostering ethical business practices is underscored, illustrating its essential integration into business strategies for a positive societal and environmental impact. The research also delves into the transition toward a circular economy, emphasizing waste management and sustainability. Moreover, the study discusses how technology integration in the financial domain has facilitated the growth of technology-centric investment platforms, aligning with sustainability criteria to attract investors who value social and environmental responsibility. Through a comprehensive analysis, this study not only demonstrates the profound impact of STI on enhancing business operations but also addresses the limitations encountered, providing a foundation for future academic pursuits in this rapidly evolving field.

The seventh chapter explains the advances toward agri-food sustainability in the European Union and its challenges and perspectives. This study undertakes a comprehensive analysis of the sustainability of the agri-food sector within the member states of the European Union (EU), addressing the environmental, social, and economic dimensions. The findings indicate moderate overall levels of sustainability in the EU agri-food sector, accompanied by notable disparities among the member states. France, Austria, and Italy are identified as leaders in sustainability efforts, whereas Malta, Cyprus, Bulgaria, and Hungary appear to be lagging. Furthermore, the study illuminates a decline in agricultural profitability and disparities in social sustainability and employment dependency, alongside the critical condition of poverty in rural regions. Environmental repercussions, such as pesticide usage, energy consumption, and organic farming practices, are also evaluated, revealing diverse sustainability practices and impacts across different countries. The document concludes by advocating for targeted national interventions to bridge sustainability gaps and augment the overall sustainability of the EU agri-food sector. It emphasizes the paramount importance of adopting innovative approaches, enhancing practices, and implementing supportive policies for farmers.

I take this opportunity to thank all the contributors for their earnest efforts to bring out this book successfully. I am sure readers of this book will find it very useful.

With best wishes,
Dr. Subramanian Senthilkannan Muthu

Reference

Beck, N., & Katz, J. N. (1995). What to Do (and Not to Do) with Time-Series Cross-Section Data. *American Political Science Review, 89*, 634–647. https://doi.org/10.2307/2082979

Environmental, Social, and Governance Sustainability Management and Financial Outcomes: Evidence from US Tech Companies

Daniel Balsalobre-Lorente, Tugba Nur, and Emre E. Topaloglu

Contents

D. Balsalobre-Lorente (✉)
Department of Applied Economics, University of Castilla La Mancha, Albacete, Spain
e-mail: daniel.balsalobre@uclm.es

T. Nur · E. E. Topaloglu
Department of Finance, Sirnak University, Sirnak, Turkiye
e-mail: tugbanur@sirnak.edu.tr; emresatopal@sirnak.edu.tr

S. S. Muthu (ed.), *Assessment of Social Sustainability Management in Various Sectors*, Environmental Footprints and Eco-design of Products and Processes,
https://doi.org/10.1007/978-3-031-81029-9_2

1 Introduction

Organizations face immense pressure to protect the interests of stakeholders and the environment due to ecological challenges like global warming, climate change, pollution, and emissions (Rahman et al., 2023). This process has led to a rise in adopting sustainable supply chain practices that help reduce waste and emissions while addressing societal concerns. By adopting sustainable supply chain practices, economies can gain a competitive edge in the market while reducing negative environmental impacts (Rajesh & Rajendran, 2020; Rajesh, 2020; Bissoondoyal-Bheenick et al., 2023). Measuring the sustainability performance of supply chains has become increasingly important for companies in terms of profitability and competitive advantage. While the primary goal is to increase the value generated in the supply chain, investors are also interested in nonfinancial values such as ESG issues (Rajesh & Rajendran, 2020). Today, the importance of ESG is increasing, and it is seen as an evaluation tool for socially responsible investment decisions (Cho, 2022). The United Nations Principles for Responsible Investment (PRI) report 2006 introduced the concept of ESG, which includes a company's environmental, social, and economic performance. In the past, firms were evaluated using quantitative indicators like financial statements, but nonfinancial values like ESG are becoming more critical (Rahman et al., 2023).

ESG is a measure used to assess a company's environmental, social, and governance practices. It combines these three areas to provide an overall measure of performance. Environmental performance measures a company's resource use, emissions reduction, and efforts to combat climate change. On the other hand, social performance refers to factors such as human rights, equality, and contribution to society. Corporate governance performance evaluates directors' duties, shareholder rights, transparency, and corporate information (Atan et al., 2018; Bahadori et al., 2021).

ESG management that highlights corporate social responsibility is becoming increasingly popular worldwide. The EU has led sustainable finance reforms and is transforming companies' operations (Zahid et al., 2022). This has resulted in the rapid development of ESG activities. Developing ESG ratings is crucial for creating a socially responsible business environment (Khan, 2022). ESG can help organizations improve economic performance through public accountability and corporate social responsibility (CSR) (Ahmad et al., 2024). Corporate social responsibility is fair behavior considering business operation's environmental, social, and economic effects. It can help businesses align their strategies with social needs and create a positive corporate image (Taliento et al., 2019). Various initiatives are being taken to integrate ESG processes into businesses. The business sector's value creation activities are aligning with Sustainable Development Goals through significant reports integration (Khan, 2022).

The traditional understanding of corporate governance views ESG spending as allocating corporate resources to activities that are not in shareholders' best interests. However, stakeholder theory emphasizes the social role played by the company. It suggests that ESG activities should not be limited to shareholder value but should be broadened to include stakeholder value (Bissoondoyal-Bheenick et al., 2023). Despite the ongoing debate in the literature, empirical evidence shows that environmental, social, and corporate sustainability indicators positively affect company performance, according to studies by Maji and Lohia (2023), Nguyen et al. (2022), and Cho (2022). However, other studies by Ruan and Liu (2021) and Al Hawaj and Buallay (2022) suggest adverse outcomes. Ruan and Liu (2021) explain that this disparity can be attributed to the following:

1. There needs to be comparability in ESG disclosure standards, scope, and focus across countries and sectors.
2. The impact of disclosing ESG information differs between environmentally sensitive and nonsensitive industries.
3. The neglect of corporate social responsibility in developing countries as opposed to developed countries depends on their level of development.

The purpose of this study is to investigate how the total ESG score and the environmental (E), social (S), and governance (G) scores of US technology companies impact their financial performance. This study aims to contribute to existing literature in two ways. Firstly, it aims to assess the impact of ESG on financial performance by considering the overall score and each ESG component separately to draw broad policy implications. Previous studies suggest that ESG components have different effects on company performance (Maji & Lohia, 2023; Nguyen et al., 2022; Cho, 2022; Ruan & Liu, 2021; Al Hawaj & Buallay, 2022) and that the findings vary across samples and time. Therefore, it is essential to reexamine these relationships. Secondly, Egorova et al. (2022) suggest that technology companies are exposed to indirect environmental risks and may face litigation, particularly antitrust litigation, in the intellectual property-focused segments of the tech industry, such as software applications, hardware devices, and semiconductor designs. Additionally, they may encounter social risks related to supply chain management, information privacy and security, human resources, and diversity. Therefore, examining the connection between ESG management and performance of technology companies can aid company managers, investors, and stakeholders in evaluating their ESG awareness.

The following is the structure of the paper: Sect. 2 presents the theoretical background of the topic. Section 3 provides a comprehensive literature review and presents the empirical results. Section 4 describes the dataset, hypotheses, and methodology used. Section 5 displays the results. Finally, Sect. 6 provides the study's conclusions and recommendations.

2 Theoretical Background

2.1 Stakeholder Theory

In 1970, Friedman argued that a corporate manager, acting as an agent, should prioritize maximizing shareholder wealth. At that time, ESG activities were considered a fiduciary problem that could distort wealth. The agency view held that companies should not allocate corporate resources to environmental, social, and governance activities because it would decrease shareholder value. When managers invest in social activities, they face agency costs and lose sight of their core governance responsibilities (Peng & Isa, 2020). Another viewpoint that differs from the traditional shareholder perspective is Freeman's (1984) stakeholder theory. According to this theory, value creation is the primary driver of business. It also acknowledges that this value should be shared with shareholders, managers, and all members of society who may be interested in how the company operates.

It is recognized that profit is important but just one outcome of the value creation process (Theodoulidis et al., 2017). In line with this view, a company's responsibility extends beyond creating shareholder value. It is reinforced and safeguarded when a company takes care of its social and environmental surroundings. This perspective has led researchers to believe that the business model must be redefined comprehensively to generate value for all stakeholders, including people and the rest of nature (Whitelock, 2019).

Stakeholder theory is a concept that focuses on the relationship between a company and all its stakeholders to ensure that they are all equally considered and satisfied. This approach aims to create sustainable value in the long term. On the other hand, the CSR model mainly focuses on promoting social responsibility, including education, healthcare, environmental conditions, philanthropy, and ethical business practices (Taliento et al., 2019). Although stakeholder theory and CSR have some similarities, they also have some differences, as stated by Freeman and Dmytriyev (2017). For example, CSR only focuses on a single set of business responsibilities (such as responsibilities to local communities and society) to ensure that companies fulfil them.

In contrast, stakeholder theory aims to understand the broader range of responsibilities in the business world. It suggests that corporate ESG efforts should be integrated into the goal of maximizing corporate value to incorporate CSR considerations. Companies focusing on ESG initiatives can gain social capital and build collaborative networks while encouraging technological innovation and efficient resource utilization (Hu et al., 2023). When companies achieve their ESG goals, they can better protect their interests and improve ESG ratings. This may lead managers to prioritize long-term corporate development over medium- and short-term performance (Tang, 2022).

2.2 *Resource-based Theory*

The resource-based view theory suggests that a company's competitive advantage is determined by its resources and capabilities. According to Barney's (1991) framework, the resources and capabilities that are valuable (i.e., those that add value to customers), rare (i.e., those that competitors cannot quickly obtain), inimitable (i.e., those that cannot be imitated or copied by competitors), and feasible (i.e., the ability to manage resources and capabilities effectively) determine a company's competitive advantage. Grant (1991) defines resources as tangible and intangible. Tangible resources include financial reserves and physical assets such as equipment, raw material inventories, and plants. On the other hand, intangible resources include human resources such as reputation, technology, organizational culture, employee learning and development, commitment, and loyalty (Sharma et al., 2019). Some proponents of the resource-based view distinguish between resources fully endorsed by the company, such as physical capital or brand names, and less tangible assets, like organizational routines and capabilities.

Similarly, there is a distinction between static and dynamic resources. Regardless of the nature of the resources, the critical requirement of the resource-based view is that they are specific to the company and intricate to imitate. Therefore, these resources enable companies to achieve long-term competitive advantage (Lockett et al., 2009). A new theoretical framework, that is, the natural resource-based view, has emerged in line with the growing interest in CSR issues. This view argues that developing competitive advantage can include managing environmental and social issues arising from CSR decisions (Battisti et al., 2022). Consequently, from the resource-based perspective, ESG activities are strategic investments. Effective management of ESG factors can result in sustainable competitive advantage because improved environmental and social performance can enhance reputation, customer satisfaction, and long-term company value.

2.3 *Sustainability Performance Indicator of ESG Scores*

There is a growing demand for companies to treat all stakeholders equally, including employees, customers, and shareholders. This is especially important given the increasing global awareness of climate-smart investing and carbon controls. To achieve this, there is a need for greater transparency in nonfinancial reporting on ESG factors (Bissoondoyal-Bheenick et al., 2023). ESG has become an essential indicator of corporate governance, risk management, and nonfinancial performance. It encompasses various issues related to the environment (climate change, use of natural resources, carbon emissions), social responsibility (occupational health and safety, workplace culture, community involvement, human rights), and corporate governance (ethics and transparency, shareholder rights, governance structure). As a result, ESG is now considered a part of CSR (Cucari et al., 2018). Carnini Pulino et al. (2022) state that the ESG score assesses a company's ESG performance. It considers the company's CSR activities and corporate governance practices objectively and transparently. Recently, there has been a shift in focus from short-term profit maximization to long-term sustainability. This has led to a debate among executives, directors, and investors about incorporating ESG factors into investment and strategic decisions. It is widely believed that ESG is an essential source of corporate risk and could impact company performance (Boze et al., 2019; Rajesh, 2020).

Various organizations assess ESG performance. However, there are differences in these organization's rating methodology's sub-criteria. This study employs Refinitiv's ESG scores, which combine three category scores into categories that make up the overall ESG score. The ESG score represents a company's ESG performance, commitment, and effectiveness. The ESG score comprises ten categories, including three subcategories for the environmental score, four for the social score, and three for the governance score (Refinitiv, 2022). See Fig. 1 for more information.

The 2022 Refinitiv report provides a detailed explanation of the various components and subcategories, outlining the specifics of each.

1. *Environmental*
   (a) Resource utilization score: Supply chain management can help a company find efficient environmental solutions and reduce material, energy, and water use.
   (b) Emissions score: This statement highlights companies' dedication and efficiency in decreasing environmental emissions.
   (c) Innovation score: Companies can create new markets by reducing environmental costs and customer burdens through new environmental technologies and processes.

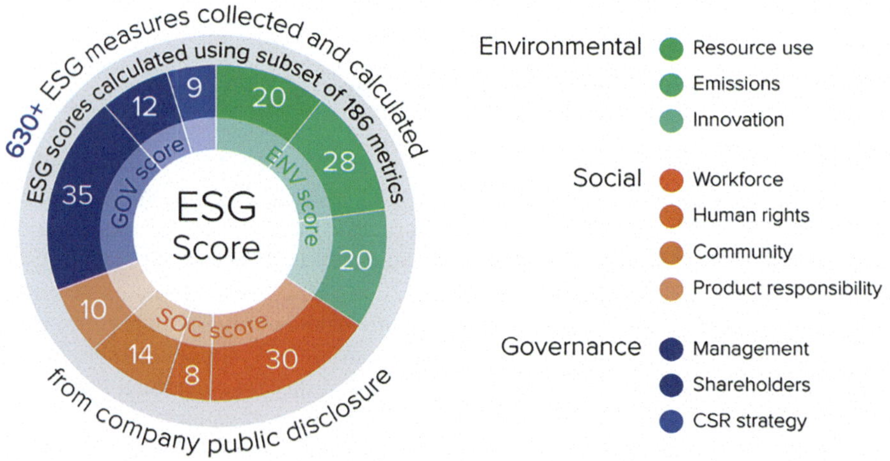

Fig. 1 ESG measures and indicators. (Source: Refinitiv, 2022)

2. *Social*
   (a) Workforce score: It measures how well an organization provides job satisfaction, job security, diversity, and equal opportunity to its employees.
   (b) Human rights score: This statement measures the effectiveness of a company in respecting fundamental human rights.
   (c) Community score: This phrase measures a company's dedication to safeguarding public health and maintaining ethical business practices.
   (d) Product responsibility score: Integrating customer privacy, health, and safety signals the company's commitment to producing quality products and services.
3. *Governance*
   (a) Governance score: It reflects the company's adoption of corporate governance principles and its commitment to and effectiveness in governance.
   (b) Shareholder score: It reflects the company's ability to treat shareholders equitably and use anti-takeover tools.
   (c) CSR strategy score: It reflects the company's practices in integrating economic, social, and environmental considerations into daily decision-making.

ESG scores are an evaluation tool used by the financial community to assess companies based on three main criteria: environmental, social, and governance. The popularity of ESG scores has risen because responsible companies tend to perform better in the long term. As a result, responsible investment strategies have produced better investment returns in previous years. High ESG scores represent less risk as a long-term investment and can improve a company's financial performance (Clement et al., 2022). However, according to Lee et al. (2023), differences in the scope and measurement of ESG ratings by companies can create confusion in investors' minds, reducing the likelihood of a clear and direct relationship with company performance.

3 Literature Review

In recent years, much attention has been given to the relationship between ESG and company performance. ESG extends corporate social responsibility and socially responsible investing (Chen & Xie, 2022). However, there are conflicting views on the impact of ESG on financial performance. Some studies have examined this relationship under stakeholder theory (Peng & Isa, 2020; Kumar, 2023; Hu et al., 2023), agency theory (Peng & Isa, 2020; Bonham & Riggs-Cragun, 2023), and resource-based theory (Sharma et al., 2019; Bhandari et al., 2022). While Rahman et al. (2023) argue that investors consider ESG to be essential for generating long-term financial benefits and attribute value to companies that improve social and environmental well-being, differences in ESG ratings may create uncertainty for investors and reduce the likelihood of a direct relationship with company performance (Lee et al., 2023). Therefore, whether ESG efforts are essential for improving financial performance or are just used for corporate image is still being determined. Some empirical studies have shown the relationship between ESG and company performance. For example, Buallay (2019) found a significant positive impact of ESG initiatives on company performance in a sample of 235 banks for the period 2007–2016, while Ting et al. (2019) and Huang (2021) found a positive relationship between ESG and company financial performance, but this relationship is economically modest. They also find that ESG activities are unlikely to be primarily motivated by narrow measures of corporate financial performance.

Several recent studies have analyzed the relationship between ESG activities and firm performance in various countries. Ruan and Liu (2021) found a significant negative relationship between ESG and company performance for companies operating in China. The authors explained this using the traditional neoclassical approach, which suggests that investing in ESG activities leads to additional cost burdens that negatively affect financial performance. On the other hand, Carnini Pulino et al. (2022), who conducted a study on Italian companies between 2011 and 2020, found that the total ESG score positively affects financial performance, while the environmental score has a negative effect. They explained the negative effect of the environmental score on financial performance by the fact that the capital invested in business activities to reduce emissions increases while the revenue generated is lower. Similarly, Maji and Lohia (2023) examined Indian companies, Fuadah et al. (2022) studied Indonesian companies, Nguyen et al. (2022) analyzed S&P500 companies, Rahman et al. (2023) focused on Pakistani companies, and Cho (2022) examined Korean companies. All of these studies found that ESG activities positively impact company performance. Lastly, Pu (2022) found an inverted U-shaped relationship between ESG activities and company performance in Chinese companies. The author explained this finding by stating that stakeholders perceive ESG activities as unnecessary, avoidable, and costly after a certain point.

Recent research studies found different relationships between ESG activities and company performance. Qu and Zhang (2023) studied companies operating in China and found a positive relationship between ESG and company performance. They also discovered that this relationship depends on the life cycle of companies. Specifically, the relationship between ESG and company performance is more evident for companies in the growth and maturity stages. Jung and Yoo (2023) analyzed 52 countries and found a linear relationship between ESG activities and company performance. However, they also found an inverted U-shaped relationship between ESG activities and company value. In contrast, Narula et al. (2024) studied Indian companies during the COVID period and found no relationship between ESG activities and company performance. The authors explained this by the fact that India, as an emerging market, has many ESG investments that lead to a decline in firm performance. Prabhakar and Vijayakumar (2024) studied 167 companies listed on the National Stock Exchange (NSE) and found that ESG activities positively impact financial performance. The authors concluded that companies should consider sustainability activities to improve their financial performance.

After analyzing the available data, it can be concluded that there is no explicit agreement on the link between ESG and a company's performance. Although most of the findings suggest that a higher ESG score positively affects the company's performance, the impact of individual ESG components is more intricate. Additionally, there are differences in results between developing and developed countries and across various industries. The differences in the scope and measurement of ESG ratings also contribute to the variability of the findings. In light of these uncertainties, it would be beneficial to investigate these relationships further.

4 Empirical Research

4.1 Data Set and Descriptive Statistics

This study investigates the relationship between ESG sustainability management and the financial performance of technology companies in the United States. The study uses data from 2017 to 2022 for seven US technology companies (ACI Worldwide Inc., Cantaloupe Inc., CoreCard Corp., Fidelity National Information Services Inc., Intuit Inc., Priority Technology Holdings Inc., and StoneCo Ltd.). The ESG sustainability scores, financial performance, company size, and age data are sourced from the Refinitiv (2024) database. The primary focus of the research is to explore the impact of the total ESG score and its components separately on the financial performance of these companies. The detailed findings of this study carry significant policy implications for the strategic objectives of managers and stakeholders. Moreover, the study includes company age and size as control variables. The literature suggests that larger companies tend to invest more in ESG sustainability as they have more resources and that older companies are more concerned with their corporate image and are more involved in sustainability initiatives (Abdi et al., 2022). Therefore, considering these variables' direct impact on ESG and company performance, they are included in the model as control variables. Table 1 explains the dependent, explanatory, and control variables used in this study. The study's explanatory variables, which are analyzed to determine their impact on financial performance, are presented in Fig. 2.

4.2 Economic Modelling

This study analyzes the link between ESG sustainability management and financial performance in US technology companies. Sustainability management uses the ESG and component scores, whereas financial performance is measured using the return on assets ratio. Additionally, the study considers the company's size, which might affect the financial performance, and measures it by the logarithm of total

Table 1 Describing variables and data sources

	Variables	Acronym	Brief
Dependent	Return on asset	ROA	Profitability relative to asset
Independent	Environmental, social, and governance	ESG	Environmental, social, and governance management
Independent	Environmental pillar score	EPS	Assesses the company's environmental performance
Independent	Social pillar score	SPS	Quantifies the company's social performance
Independent	Governance pillar score	GPS	Evaluate the company's governance practices
Control	Company size	SIZE	Natural logarithm of total assets
Control	Date of Incorporation	AGE	Company age is the year a company officially becomes an initial public offering

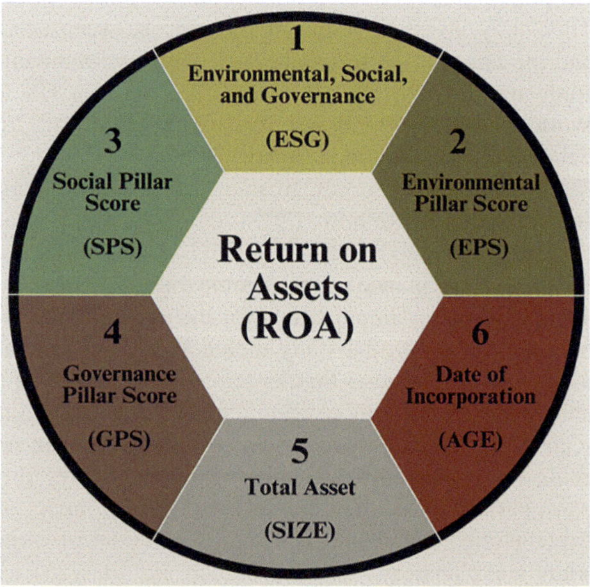

Fig. 2 Framework of the research

assets. The company's age is also measured by the years since its initial public offering. Equation 1 presents the financial performance function.

$$ROA_{it} = f\left(ESG_{it}, EPS_{it}, SPS_{it}, GPS_{it}, SIZE_{it}, AGE_{it}\right) \quad (1)$$

In estimating the model using the research function, the logarithmic form is applied to the SIZE and AGE series. In contrast, the ROA, ESG, EPS, SPS, and GPS series are analyzed proportionally. Equation 2 reports the estimated econometric model.

$$ROA_{it} = \alpha_0 + \alpha_1 ESG_{it} + \alpha_2 EPS_{it} + \alpha_3 SPS_{it} + \alpha_4 GPS_{it} + \alpha_5 LnSIZE_{it} + \alpha_6 LnAGE_{it} + \mu_{it} \quad (2)$$

In Eq. 2, the constant term α_0 represents the estimation parameters $\alpha_1 - \alpha_6$, μ_{it} represents the error term, *t* and *i* represent the 2017–2022 period and US technology companies in the panel data.

4.3 Methodology

In this study, we have constructed a methodological framework for analyzing panel data of US technology companies in line with time series data. The first step is calculating descriptive statistics for the variables and checking for normal distribution. We then create correlation matrices and analyze problems such as multicollinearity and endogeneity. Since the time dimension is narrow, panel data does not require the stationarity process. We do not apply cross-sectional dependence, slope heterogeneity, and unit root tests on a variable basis. Instead, we analyze the presence of period and cross-sectional effects in the model to be estimated. We then estimate elasticity coefficients using the fixed effects model when analyzing a specific group. After that, we analyze problems such as cross-sectional dependence, serial correlation, and heteroscedasticity in the fixed effects model. Finally, we determine elasticity coefficients using the robust estimation method that solves the standard errors of CSD, autocorrelation, and variance in panel data. See Fig. 3 for the research strategy.

4.3.1 Multicollinearity and Endogeneity Evidence

Multicollinearity is used in panel regression models to describe a strong or nearly perfect linear relationship between some or all the explanatory variables. It becomes problematic when the correlation between the error term of the model and the explanatory variables causes an endogeneity issue. This high correlation between the variables can make calculating the parameters difficult or even impossible, rendering the least squares method useless. Therefore, as part of panel data analysis, Spearman's correlation analysis is essential to determine any issues with multicollinearity and endogeneity. Under normal distribution conditions, Spearman's correlation test, as shown in Eq. 3, can construct the correlation matrix.

$$r_s = 1 - \frac{6\sum_{i=1}^{n} d_i^2}{n\left(n^2 - 1\right)} \quad (3)$$

In the abovementioned equation, $i = 1,2,...,n$, (xi, yi) represents the ith measurement value of the variables. Once we have confirmed that the model does not suffer from endogeneity or multicollinearity issues, we can confidently state that the estimation model is accurate and reliable.

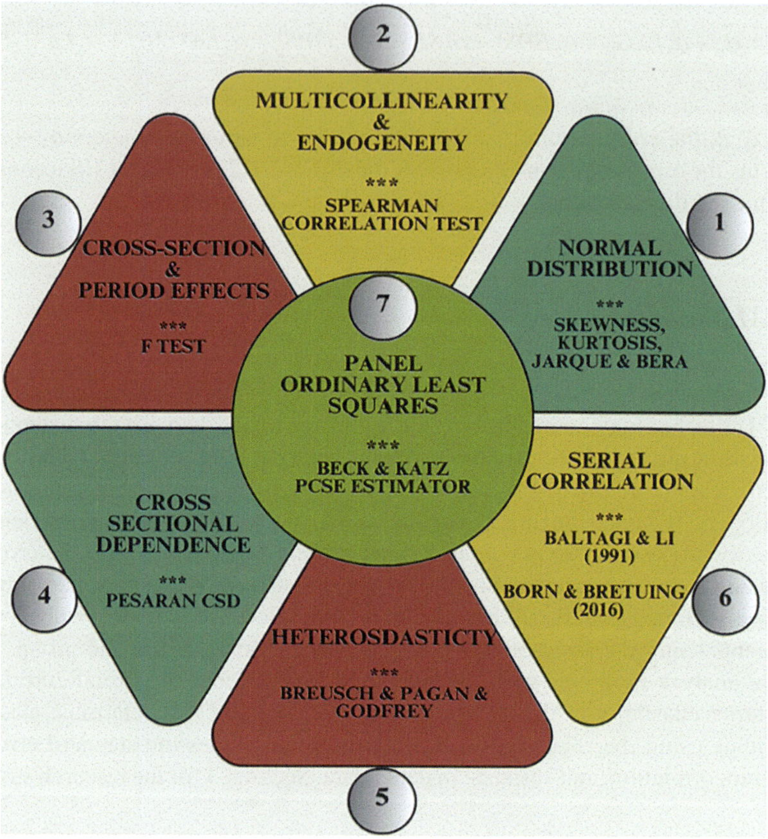

Fig. 3 Flowchart of the research method

4.3.2 Determination of the Estimation Model (Cross-sectional and Period Effects)

In order to assess the effect of ESG sustainability variables on the financial performance of US technology companies, it is critical to analyze period and company effects in panel data. This study focuses on data for a specific period and industry operating within a specific country. The F-test is used to assess the validity of the fixed effects model to determine the best panel data model to represent the data. This test looks for company-specific unit effects and determines whether the classical pooled or fixed effects model is better. Two types of models are used, namely, restricted and unrestricted models. In the unrestricted model, the data on the variables take values according to the units, while in the restricted model, unit differences are not considered necessary. The constrained and unconstrained models are represented by Eqs. 4 and 5, respectively.

$$\text{Restricted model}: Y_i = X\beta + u \tag{4}$$

$$\text{Unrestricted model}: Y_i = X_i \beta_i + u_i \tag{5}$$

Equations 4 and 5 test the null hypothesis $H_0 : \beta_i = \beta$ (no cross-sectional and period effects). If the null hypothesis is rejected, then the fixed effects ordinary least-squares (OLS) model is applied.

4.3.3 Cross-sectional Dependency

Various factors can trigger cross-sectional dependence (CSD) in technology companies based in the United States, including merging financial and economic factors. If one company experiences a sudden shock, it can affect other companies in the same industry. Therefore, it is crucial to consider CSD while analyzing panel data, as ignoring it can lead to inaccurate and unreliable results (Breusch & Pagan, 1980; Pesaran, 2004). Failure to account for CSD can result in biased and inconsistent analysis outcomes. Hence, testing for cross-sectional dependence in the series is essential before analyzing the panel data. The Pesaran (2004) test is used to check for the presence of cross-sectional dependence among series in panel data with a larger cross-sectional dimension than the time dimension ($N > T$). Equation 6 is utilized to analyze CSD.

$$\text{CSD} = \sqrt{\frac{2\text{T}}{N(N-1)}} \left(\sum_{\text{i=1}}^{\text{N-1}} \sum_{\text{j=i+1}}^{\text{N}} \hat{\rho} ij \right) N(0,1) i, j \tag{6}$$

The abovementioned equation tests the null hypothesis stated in Eq. 7.

$$H_0 : \rho_{ij} = \text{cov}\left(\mu_{it}, \mu_{jt}\right) = 0, \tag{7}$$

If the null hypothesis cannot be rejected, the presence of CSD in the model is inferred.

4.3.4 Testing Heteroskedasticity and Serial Correlation

The basic assumptions about the model error term in panel data analysis are heteroskedasticity and autocorrelation. When we say heteroskedasticity, it means that the assumption of constant variance is not valid. This means that the variances of the error terms are different for each cross section, and their covariances are not 0. In other words, there is a variation in the conditional variance of the error term. This is a common problem, especially when analyzing cross-sectional data.

The autocorrelation assumption in panel data analysis refers to the relationship between successive values of the error term. This means that unit values are not independent and can affect one another, leading to a systematic relationship. This situation can cause biases and inconsistencies in the analysis. Autocorrelation is a common problem in panel data analysis that involves analyzing time and cross-sectional dimensions. To test for heteroscedasticity, we use the heteroscedasticity

Lagrangian multiplier (LM) test of Breusch and Pagan (1979), while the tests of Baltagi and Li (1991) and Born and Bretuing (2016) are used to analyze autocorrelation.

4.3.5 Elasticity Coefficient Estimation Method

Panel data regression estimation can encounter standard errors such as cross-sectional correlation, nonconstant variance of the model error term, and nonindependence of successive values of the error term. Beck and Katz (1995, 1996) proposed a simplified method to address this problem. They retain the OLS parameter estimates, which are consistent but inefficient, while replacing the OLS standard errors with panel-corrected standard errors (PCSE). Beck and Katz claim that OLS with PCSE is superior to the Kmenta FGLS approach when estimating small sample panel models. Beck and Katz (1995, 1996) conducted Monte-Carlo experiments and found that PCSE are more efficient than OLS standard errors. PCSE can be used with OLS or Prais-Winsten (Prais & Winsten, 1954) regressions. Additionally, PCSE extend White's (1980) heteroskedasticity-consistent standard errors and take into account the panel structure of the data. As a result, PCSE outperform White's variance-consistent standard errors for panel data.

This study aims to analyze the link between the ESG sustainability scores and the financial performance of US technology companies. The PCSE estimator, which considers standard errors in the panel data model and provides reliable outcomes for narrow panel data, is used for the analysis.

5 Results

This study aims to analyze the financial performance of technology companies in the United States in terms of their environmental, social, and governance sustainability management. The research employs the panel data analysis method, where the first step involves computing descriptive statistics for the series. Table 2 presents the mean, standard deviation, skewness, and kurtosis coefficients indicating normal distribution, along with the Jarque-Bera (JB) test statistics for each series. Furthermore, box charts are generated for each series set's quarterly mean, dispersion, and variability indicators and scatter plots for each model.

In the given descriptive statistics, the mean values for ROA, ESG, EPS, GPS, SPS, AGE, and SIZE series are 0.043, 43.567, 17.562, 46.521, 49.650, 3.480, and 21.171, respectively. On the other hand, the standard deviation values for ROA, ESG, EPS, GPS, SPS, AGE, and SIZE series are 0.126, 21.658, 27.166, 25.274, 26.736, 0.440, and 2.377, respectively. Among these, the highest deviation from the mean is calculated for ESG, while the lowest standard deviation is for ROA. This suggests that there are high variations, especially in sustainability management variables. Skewness, kurtosis, and JB test statistics are indicators of normal distribution.

Table 2 Descriptive statistics and correlations matrix

Box Plots

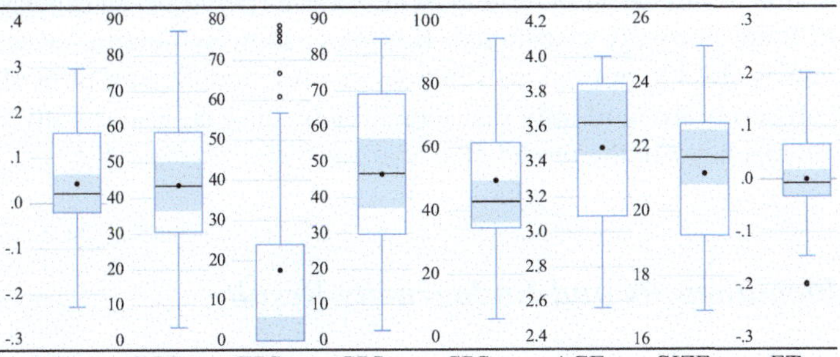

Stats.	ROA	ESG	EPS	GPS	SPS	AGE	SIZE	ET
Mean	0.043	43.567	17.562	46.521	49.650	3.480	21.171	-1E-18
Median	0.022	43.439	0.000	46.802	43.040	3.624	21.675	-0.007
Maximum	0.298	86.970	78.452	84.625	94.656	4.007	25.152	0.202
Minimum	-0.229	3.796	0.000	2.719	5.833	2.565	16.896	-0.200
Std. Dev.	0.126	21.658	27.166	25.274	26.736	0.440	2.377	0.088
Skewness	0.244	-0.094	1.262	-0.237	0.339	-0.594	-0.098	-0.135
Kurtosis	2.839	2.313	3.016	1.790	2.118	1.983	1.995	3.108
Jarque-Bera	0.461	0.888	11.149*	2.957	2.164	4.280	1.836	0.149
Observations	42	42	42	42	42	42	42	42

Scatters

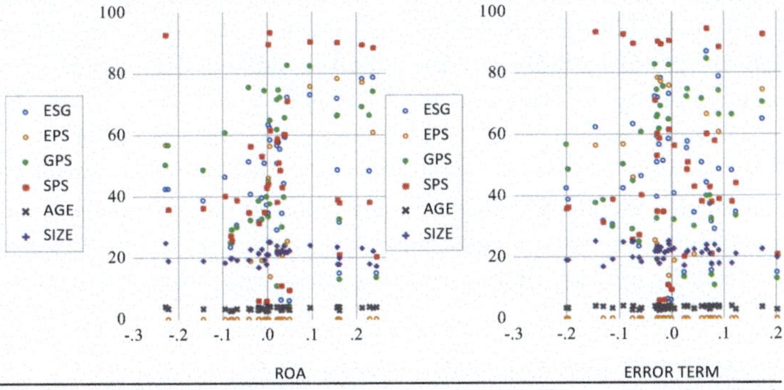

Note: ROA, ESG, EPS, GPS, SPS, AGE, SIZE, and ET denote the Return on assets, environmental, social, and governance, Environmental sustainability score, Governance sustainability score, Social sustainability score, Company age, Company size, and model's error term. * denotes the significance 1%

In a normal distribution, the skewness and kurtosis coefficients should be 0. However, for financial data, the skewness value is expected to be close to 0, and the kurtosis value is expected to be close to 3 for the normal distribution to be realized. The JB test tests the null hypothesis that the series are normally distributed. The null hypothesis must be rejected for a normal distribution to be valid. The ROA, EPS, and SPS series have positive skewness coefficients, indicating a left-skewed distribution. On the other hand, the ESG, GPS, AGE, and SIZE series have negative skewness coefficients, indicating a right-skewed distribution. All series have positive kurtosis coefficients, indicating a skewed distribution. According to the JB test results, the null hypothesis is rejected for the EPS series, while it cannot be rejected for the ROA, ESG, GPS, SPS, AGE, and SIZE series. Therefore, all other variables except for EPS are normally distributed. Box plots show symmetric effects in the series, and values close to normal distribution are observed. Scatter plots show that the data for all series are randomly distributed, indicating no issues with multicollinearity or endogeneity in the model.

5.1 Multicollinearity and Endogeneity Results

Multicollinearity is a perfect or almost perfect linear relationship between some or all of the explanatory variables in a panel data model. The strong correlation between the explanatory variables and the error term can cause an endogeneity problem in the model. The correlation coefficient exceeding 0.90 indicates multicollinearity and endogeneity problems (Tabachnick & Fidell, 2001). A high degree of correlation can make it impossible to calculate the parameters and render the least squares method unusable.

After conducting the descriptive statistics calculations, it was discovered that none of the series were usually distributed. As a result, Spearman's correlation test was used to analyze multicollinearity and endogeneity since it provides unbiased results in the absence of normal distribution. Bivariate correlation matrices for logarithmic and proportional variables are presented in Table 3. The analysis revealed a strong correlation between ROA and the explanatory and control variables. The highest correlation was found between ESG and GPS, with a coefficient of 0.890. This suggests no multicollinearity issue among the explanatory variables in the model. The highest correlation coefficient between the model's error term and the explanatory variables was also calculated as 0.104, which occurred between ET and EPS. This result indicates that there is no endogeneity problem in the model.

5.2 Estimation Model Determination Results

To analyze the impact of ESG, EPS, GPS, and SPS scores, along with SIZE and AGE control variables, on financial performance, it is necessary to examine cross-sectional and time effects in panel data. This study analyzes data from US technology companies over a specific period, and the F-test is used to determine which of the fixed effects models and the classical pooled model provides more efficient and consistent results. The F-test results show a cross-sectional effect in the estimation model, and the null hypothesis is rejected according to the calculated F-group_fixed test probability value. However, the null hypothesis cannot be rejected according to the F-time_fixed test probability value, and it is determined that there is no time effect in the estimation model (Table 4). The panel data in this study consists of

Table 3 Correlation spreadsheet for multicollinearity and endogeneity

Correlation	ROA	ET	ESG	EPS	GPS	SPS	AGE	SIZE
ROA	1.000							
ET	0.682	1.000						
ESG	0.286	−0.015	1.000					
EPS	0.305	−0.104	0.717	1.000				
GPS	0.210	0.043	0.890	0.432	1.000			
SPS	0.219	0.019	0.829	0.827	0.588	1.000		
AGE	0.312	0.035	0.475	0.293	0.377	0.524	1.000	
SIZE	0.083	−0.013	0.477	0.699	0.168	0.738	0.281	1.000
t-Statistic	ROA	ET	ESG	EPS	GPS	SPS	AGE	SIZE
ROA	-----							
ET	5.901	-----						
ESG	1.891	−0.095	-----					
EPS	2.028	−0.659	6.514	-----				
GPS	1.357	0.270	12.353	3.030	-----			
SPS	1.420	0.118	9.376	9.314	4.596	-----		
AGE	2.077	0.220	3.410	1.940	2.573	3.894	-----	
SIZE	0.526	−0.083	3.434	6.179	1.079	6.926	1.855	-----

Table 4 Cross-sectional and period effect determination results

Test	Statistics	Probability	Hypothesis	Decision
F-group	6.469*	0.000	H_0: There is a cross-sectional effect but no time effect	Reject
F-period	0.634	0.676	H_0: There is a time effect but no cross-sectional effect	Cannot be rejected
F-two way	4.143*	0.002	H_0: No cross-sectional and time effect	Reject

* denotes the significance 1%

6 years and seven companies. Therefore, companies are more effective than periods in the effect of explanatory variables on financial performance.

5.3 *Cross-sectional Dependency Test Result*

Considering the interdependence among technology companies with high environmental, social, and governance sustainability management levels is deemed essential. If we fail to do so, the analysis outcomes might be biased and inconsistent (Breusch & Pagan, 1980; Pesaran, 2004). Therefore, testing for cross-sectional dependence in the series is necessary before conducting panel data analysis. In this study, we used Pesaran's (2004) CSD test for cross-sectional dependence, and the evidence we obtained rejects the null hypothesis that there is no cross-sectional dependence in the model (Table 5). As a result, there is cross-sectional dependence in the panel data. In other words, a shock or crisis in one technology company affects the others. Therefore, to improve their financial performance, each company should also consider the shocks that affect the financial performance of other companies.

5.4 Heteroskedasticity and Serial Correlation Test Results

Panel data analysis is often faced with the problem of variation in the conditional variance of the error term due to the horizontal cross-sectional dimension. This situation creates a systematic relationship in panel data, leading to an autocorrelation problem. When estimating elasticity coefficients, it is essential to take into consideration the issues of heteroskedasticity and autocorrelation. Table 6 presents the results of the varying variance and autocorrelation tests. The Breusch-Pagan-Godfrey LM test probability value indicates that the null hypothesis of no varying variance is rejected, meaning that the variances of the error terms are not constant across all cross sections and their covariances also differ from 0. The Baltagi and Li (1991) LM and Born and Bretuing (2016) LM test probability values suggest that the null hypothesis of no autocorrelation cannot be rejected. Thus, the successive values of the error terms are not independent, indicating variance in the panel data model but no autocorrelation.

5.5 Model Estimation Results

The current study employs the PCSE method to estimate the elasticity coefficient, which is robust to autocorrelation and panel standard errors of variance. This method examines the relationship between environmental, social, and governance sustainability scores and financial performance. In the estimated panel data model, it is found that the variances of the error terms are not constant, and the successive values of the error terms are not independent. The estimation results, as reported in Table 7, indicate that the explanatory variables elucidate approximately 34% (R^2_{adj}) of the variation in financial performance. The probability value of the F-statistic, which expresses the significance of the panel data model as a whole, is significant at the 95% confidence level. These findings confirm statistically significant

Table 5 Cross-sectional dependency result

H_0: No cross-sectional dependency		
	Statistics	Probability
Model	−1.676*	0.093

* denotes the significance 10%

Table 6 Heteroskedasticity and serial correlation test results for fixed effect model

Hypothesis	Test	Statistics	Probability
H_0: No Heteroskedasticity	Breusch-Pagan-Godfrey LM	14.788*	0.022
H_0: No serial correlation	Baltagi and Li (1991) LM	0.115	0.734
	Born and Bretuing (2016) LM	2.588	0.108

* denotes the significance 5%

Table 7 PCSE FE-OLS estimation results

	Model: $ROA_{it} = f(ESG_{it}, EPS_{it}, GPS_{it}, SPS_{it}, AGE_{it}SIZE)$			
Parameters	Coefficient	Std. error	Statistics	Probability
ESG	0.013*	0.003	3.814	0.001
EPS	0.005*	0.001	4.157	0.000
GPS	−0.007*	0.002	−3.620	0.001
SPS	−0.008*	0.002	−3.501	0.002
AGE	0.134*	0.042	3.214	0.003
SIZE	−0.021**	0.009	−2.194	0.036
C	0.097	0.227	0.427	0.673
Adj. R^2	0.336**			

*, ** denotes the significance 1% and 5% respectively

relationships between all explanatory and control variables and financial performance for US technology companies. Moreover, the study finds a positive relationship between ESG and ROA. It is found that 1-unit increase in ESG increases ROA by 0.013 units. This suggests that US technology company's environmental, social, and governance sustainability management positively impacts their financial performance. Stakeholders perceive the company's ESG disclosures positively, and companies can integrate sustainability efforts into business operations. Therefore, these companies can positively influence company performance by improving nonfinancial disclosure. This finding is consistent with the findings of other studies. For instance, Ting et al. (2019) found similar results for emerging and developed market samples. Maji and Lohia (2023) also found similar results for Indian companies, Fuadah et al. (2022) for Indonesian companies, Nguyen et al. (2022) for S&P500 companies, Rahman et al. (2023) for Pakistani companies, and Cho (2022) for Korean companies.

There is a positive correlation between EPS, which measures the environmental performance of technology companies, and ROA. Specifically, a 1-unit increase in EPS leads to a 0.005-unit increase in ROA. This implies that a company's environmental sustainability and risk management activities can enhance its financial performance. On the other hand, GPS, which measures corporate governance practices, is negatively related to ROA. A 1-unit increase in GPS results in a 0.007-unit decrease in ROA. This suggests that the governance structure of technology companies needs to be improved, and their ability to mitigate risks is low. However, companies may prioritize short-term objectives over good governance practices, requiring additional costs and resources. Additionally, changes in the corporate structure to improve governance practices may need to be better received by stakeholders. Furthermore, SPS, which measures a company's social performance, is negatively associated with ROA. A 1-unit increase in SPS leads to a 0.008-unit decrease in ROA. This implies that the social responsibility activities of technology companies are inadequate and do not positively impact stakeholders, thereby reducing financial performance. Investments made to reduce environmental impact and contribute to society are costly in the short term. Therefore, companies must adopt long-term strategies to ensure social performance translates to financial performance.

The study found that the ESG sustainability score's impact on US technology company's financial performance is limited when considered separately. However, when these three factors are considered together, they positively impact financial performance. This study also found a positive correlation between the control variable AGE, representing the years a company has operated since its initial public offering, and the financial performance indicator ROA. Specifically, a unit increase in AGE leads to a 0.134 unit increase in ROA. This finding suggests that US technology company's experience, stability, and growth potential positively impact their financial performance. On the other hand, the control variable SIZE, which measures company size, was negatively related to ROA. Specifically, a 1% increase in SIZE leads to a 0.021% decrease in ROA. This finding suggests that US technology companies must use their assets more effectively and their operations could be more efficient and adequate. Furthermore, they cannot take advantage of economies of scale and scope with increasing company size, indicating poor managerial decisions. This finding supports the negative relationship between governance sustainability scores and financial performance. Overall, the study shows that US technology company's financial performance is influenced by a combination of environmental, social, and governance sustainability scores and their age and size. The results of the study are summarized in Fig. 4.

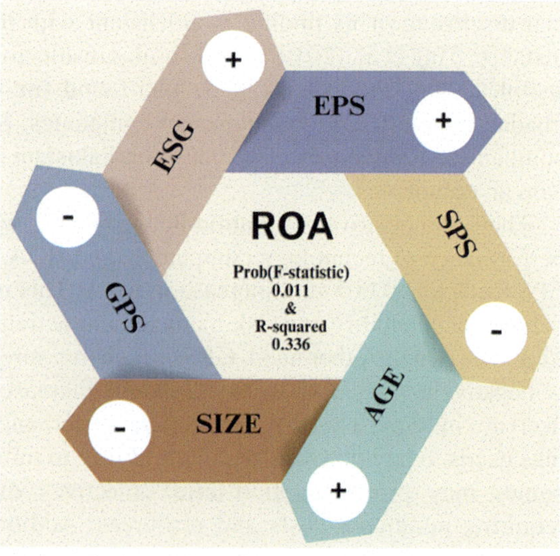

Fig. 4 Graphical summary of results for US tech companies

6 Conclusion and Policy Implication

ESG management has become a widespread corporate social responsibility function worldwide in recent years. ESG activities help companies align their strategies with societal needs and improve their corporate image. However, there are still debates on how a company's ESG activities impact company performance. This study examines the relationship between the total ESG score and ESG components and the financial performance of seven technology companies operating in the United States using panel data from 2017 to 2022. The study uses PCSE FE-OLS estimation, which considers the problem of variance and autocorrelation. The findings show a significant positive relationship between the total ESG score and financial performance. However, the relationships between ESG components and financial performance are different. There is a positive linkage between the environmental score (E) and financial performance but a negative relationship between the social score (S) and governance score (G) and financial performance. In addition, company size hurts financial performance, while company age has a positive effect.

The following policy recommendations are presented after analyzing the data and conducting discussions. The results show that the overall ESG score positively impacts financial performance. This implies that companies' disclosure of nonfinancial information leads to an improvement in their financial performance. This supports the stakeholder theory and the resource-based perspective. To enable investors and stakeholders to make better decisions and enhance the performance of these companies, it is recommended that transparent information on ESG performance be provided and related strategies prioritized. This may also encourage other companies and support sustainable development by facilitating knowledge and experience sharing among companies. Secondly, the environmental performance of these companies has a positive impact on their financial performance. Therefore, these companies' transparent reporting of environmental risks and opportunities is reflected in their financial performance. Companies should be incentivized to develop environmentally efficient solutions that reduce emissions, energy and water consumption and environmentally friendly products and services. This will improve their financial performance. Thirdly, social and governance scores negatively affect company performance in this group of companies. Social responsibility initiatives and governance practices require long-term strategies and may entail companies to expend more resources, which may conflict with the company's short-term strategies. Therefore, these companies can be encouraged to invest in social responsibility policies. For good governance practices, they can be advised to identify risks and weaknesses related to management and emphasize corporate governance scores. In this way, companies focusing on sustainable goals can improve their financial performance, benefit society, and support sustainable development goals.

References

Abdi, Y., Li, X., & Càmara-Turull, X. (2022). Exploring the impact of sustainability (ESG) disclosure on firm value and financial performance (FP) in the airline industry: The moderating role of size and age. *Environment, Development and Sustainability, 24*(4), 5052–5079. https://doi.org/10.1007/s10668-021-01649-w

Ahmad, H., Yaqub, M., & Lee, S. H. (2024). Environmental-, social-, and governance-related factors for business investment and sustainability: A scientometric review of global trends. *Environment, Development and Sustainability, 26*(2), 2965–2987. https://doi.org/10.1007/s10668-023-02921-x

Al Hawaj, A. Y., & Buallay, A. M. (2022). A worldwide sectorial analysis of sustainability reporting and its impact on firm performance. *Journal of Sustainable Finance & Investment, 12*(1), 62–86. https://doi.org/10.1080/20430795.2021.1903792

Atan, R., Alam, M. M., Said, J., & Zamri, M. (2018). The impacts of environmental, social, and governance factors on firm performance: Panel study of Malaysian companies. *Management of Environmental Quality: An International Journal, 29*(2), 182–194. https://doi.org/10.1108/MEQ-03-2017-0033

Bahadori, N., Kaymak, T., & Seraj, M. (2021). Environmental, social, and governance factors in emerging markets: The impact on firm performance. *Business Strategy & Development, 4*(4), 411–422. https://doi.org/10.1002/bsd2.167

Baltagi, B. H., & Li, Q. (1991). A joint test for serial correlation and random individual effects. *Statistics & Probability Letters, 11*(3), 277–280. https://doi.org/10.1016/0167-7152(91)90156-L

Barney, J. (1991). Firm resources and sustained competitive advantage. *Journal of Management, 17*(1), 99–120. https://doi.org/10.1177/014920639101700108

Battisti, E., Nirino, N., Leonidou, E., & Thrassou, A. (2022). Corporate venture capital and CSR performance: An extended resource-based view's perspective. *Journal of Business Research, 139*, 1058–1066. https://doi.org/10.1016/j.jbusres.2021.10.054

Beck, N., & Katz, J. N. (1995). What to do (and not to do) with time-series cross-section data. *American Political Science Review, 89*(3), 634–647. https://doi.org/10.2307/2082979

Beck, N., & Katz, J. N. (1996). Nuisance vs. substance: Specifying and estimating time-series-cross-section models. *Political Analysis, 6*, 1–36. https://doi.org/10.1093/pan/6.1.1

Bhandari, K. R., Ranta, M., & Salo, J. (2022). The resource-based view, stakeholder capitalism, ESG, and sustainable competitive advantage: The firm's embeddedness into ecology, society, and governance. *Business Strategy and the Environment, 31*(4), 1525–1537. https://doi.org/10.1002/bse.2967

Bissoondoyal-Bheenick, E., Brooks, R., & Do, H. X. (2023). ESG and firm performance: The role of size and media channels. *Economic Modelling, 121*, 106203. https://doi.org/10.1016/j.econmod.2023.106203

Bonham, J., & Riggs-Cragun, A. (2023). A different approach to agency theory and implications for ESG. *Seattle University Law Review, 47*, 635–642.

Born, B., & Breitung, J. (2016). Testing for serial correlation in fixed-effects panel data models. *Econometric Reviews, 35*(7), 1290–1316. https://doi.org/10.1080/07474938.2014.976524

Boze, B., Krivitski, M., Larcker, D. F., Tayan, B., & Zlotnicka, E. (2019). The business case for ESG. *Rock Center for Corporate Governance at Stanford University Closer Look Series: Topics, Issues and Controversies in Corporate Governance No. CGRP-77*. Available at SSRN: https://ssrn.com/abstract=3393082

Breusch, T. S., & Pagan, A. R. (1979). A simple test for heteroscedasticity and random coefficient variation. *Econometrica: Journal of the Econometric Society, 47*, 1287–1294. https://doi.org/10.2307/1911963

Breusch, T. S., & Pagan, A. R. (1980). The lagrange multiplier test and its applications to model specification in econometrics. *The Review of Economic Studies, 47*(1), 239–253. https://doi.org/10.2307/2297111

Buallay, A. (2019). Is sustainability reporting (ESG) associated with performance? Evidence from the European banking sector. *Management of Environmental Quality: An International Journal, 30*(1), 98–115. https://doi.org/10.1108/MEQ-12-2017-0149

Carnini Pulino, S., Ciaburri, M., Magnanelli, B. S., & Nasta, L. (2022). Does ESG disclosure influence firm performance? *Sustainability, 14*(13), 7595. https://doi.org/10.3390/su14137595

Chen, Z., & Xie, G. (2022). ESG disclosure and financial performance: Moderating role of ESG investors. *International Review of Financial Analysis, 83*, 102291. https://doi.org/10.1016/j.irfa.2022.102291

Cho, Y. (2022). ESG and firm performance: Focusing on the environmental strategy. *Sustainability, 14*(13), 7857. https://doi.org/10.3390/su14137857

Clément, A., Robinot, É., & Trespeuch, L. (2022). Improving ESG scores with sustainability concepts. *Sustainability, 14*(20), 13154. https://doi.org/10.3390/su142013154

Cucari, N., Esposito De Falco, S., & Orlando, B. (2018). Diversity of board of directors and environmental social governance: Evidence from Italian listed companies. *Corporate Social Responsibility and Environmental Management, 25*(3), 250–266. https://doi.org/10.1002/csr.1452

Egorova, A. A., Grishunin, S. V., & Karminsky, A. M. (2022). The impact of ESG factors on the performance of information technology companies. *Procedia Computer Science, 199*, 339–345. https://doi.org/10.1016/j.procs.2022.01.041

Freeman, R. E. (1984). *Strategic management: A stakeholder approach*. Pitman Publishing.

Freeman, R. E., & Dmytriyev, S. (2017). Corporate social responsibility and stakeholder theory: Learning from each other. *Symphonya. Emerging Issues in Management, 1*, 7–15. https://doi.org/10.4468/2017.1.02freeman.dmytriyev

Friedman, M. (1970). The social responsibility of business is to increase its profits. *New York Time Magazine*, 32–33. https://nyti.ms/1LSi5ZD

Fuadah, L. L., Mukhtaruddin, M., Andriana, I., & Arisman, A. (2022). The ownership structure, and the environmental, social, and governance (ESG) disclosure, firm value and firm performance: The audit committee as moderating variable. *Economies, 10*(12), 314. https://doi.org/10.3390/economies10120314

Grant, R. M. (1991). The resource-based theory of competitive advantage: Implications for strategy formulation. *California Management Review, 33*(3), 114–135. https://doi.org/10.2307/41166664

Hu, A., Yuan, X., Fan, S., & Wang, S. (2023). The Impact and mechanism of corporate ESG construction on the efficiency of regional green economy: An empirical analysis based on signal transmission theory and stakeholder theory. *Sustainability, 15*(17), 13236. https://doi.org/10.3390/su151713236

Huang, D. Z. (2021). Environmental, social and governance (ESG) activity and firm performance: A review and consolidation. *Accounting & Finance, 61*(1), 335–360. https://doi.org/10.1111/acfi.12569

Jung, Y. L., & Yoo, H. S. (2023). Environmental, social, and governance activities and firm performance: Global evidence and the moderating effect of market competition. *Corporate Social Responsibility and Environmental Management, 30*(6), 2830–2839. https://doi.org/10.1002/csr.2518

Khan, M. A. (2022). ESG disclosure and firm performance: A bibliometric and meta analysis. *Research in International Business and Finance, 61*, 101668. https://doi.org/10.1016/j.ribaf.2022.101668

Kumar, S. (2023). A review ESG performance as a measure of stakeholders theory. *Academy of Marketing Studies Journal, 27*(S3), 1–18.

Lee, M. T., Raschke, R. L., & Krishen, A. S. (2023). Understanding ESG scores and firm performance: are high-performing firms E, S, and G-balanced? *Technological Forecasting and Social Change, 195*, 122779. https://doi.org/10.1016/j.techfore.2023.122779

Lockett, A., Thompson, S., & Morgenstern, U. (2009). The development of the resource-based view of the firm: A critical appraisal. *International Journal of Management Reviews, 11*(1), 9–28. https://doi.org/10.1111/j.1468-2370.2008.00252.x

Maji, S. G., & Lohia, P. (2023). Environmental, social and governance (ESG) performance and firm performance in India. *Society and Business Review, 18*(1), 175–194. https://doi.org/10.1108/SBR-06-2022-0162

Narula, R., Rao, P., Kumar, S., & Matta, R. (2024). ESG scores and firm performance-evidence from emerging market. *International Review of Economics & Finance, 89*, 1170–1184. https://doi.org/10.1016/j.iref.2023.08.024

Nguyen, D. T., Hoang, T. G., & Tran, H. G. (2022). Help or hurt? The impact of ESG on firm performance in S&P 500 non-financial firms. *Australasian Accounting, Business and Finance Journal, 16*(2), 91–102. https://doi.org/10.14453/aabfj.v16i2.7

Peng, L. S., & Isa, M. (2020). Environmental, social and governance (ESG) practices and performance in Shariah firms: Agency or stakeholder theory? *Asian Academy of Management Journal of Accounting & Finance, 16*(1), 10.21315/aamjaf2020.16.1.1.

Pesaran, M. H. (2004). General diagnostic tests for cross section dependence in panels. Cambridge Working Papers.

Prabhagar, M. V., & Vijayakumar, N. (2024). *Does ESG disclosure really influence the firm performance?* The Quarterly Review of Economics and Finance. https://doi.org/10.1016/j.qref.2024.03.008

Prais, S. J., & Winsten, C. B. (1954). *Trend estimators and serial correlation*. Chicago: Cowles Commission discussion paper, 383, 1–26.

Qu, W., & Zhang, J. (2023). Environmental, social, and corporate governance (ESG), life cycle, and firm performance: Evidence from China. *Sustainability, 15*(18), 14011. https://doi.org/10.3390/su151814011

Rahman, H. U., Zahid, M., & Al-Faryan, M. A. S. (2023). ESG and firm performance: The rarely explored moderation of sustainability strategy and top management commitment. *Journal of Cleaner Production, 404*, 136859. https://doi.org/10.1016/j.jclepro.2023.136859

Rajesh, R. (2020). Exploring the sustainability performances of firms using environmental, social, and governance scores. *Journal of Cleaner Production, 247*, 119600. https://doi.org/10.1016/j.jclepro.2019.119600

Rajesh, R., & Rajendran, C. (2020). Relating environmental, social, and governance scores and sustainability performances of firms: An empirical analysis. *Business Strategy and the Environment, 29*(3), 1247–1267. https://doi.org/10.1002/bse.2429

Refinitiv (2022). Environmental, Social and Governance (ESG) Scores from Refinitiv- May 2022, Refinitiv, an LSEG business.

Refinitiv (2024). Environmental, Social and Governance Scores, Brochure on Refinitiv Website, April 2024.

Ruan, L., & Liu, H. (2021). Environmental, social, governance activities and firm performance: Evidence from China. *Sustainability, 13*(2), 767. https://doi.org/10.3390/su13020767

Sharma, D., Bhattacharya, S., & Thukral, S. (2019). Resource-based view on corporate sustainable financial reporting and firm performance: Evidences from emerging Indian economy. *International Journal of Business Governance and Ethics, 13*(4), 323–344. https://doi.org/10.1504/IJBGE.2019.099565

Tabachnick, B. G., & Fidell, L. S. (2001). *Using multivariate statistics*. Allyn and Bacon.

Taliento, M., Favino, C., & Netti, A. (2019). Impact of environmental, social, and governance information on economic performance: Evidence of a corporate' sustainability advantage' from Europe. *Sustainability, 11*(6), 1738. https://doi.org/10.3390/su11061738

Tang, H. (2022). The effect of ESG performance on corporate innovation in China: The mediating role of financial constraints and agency cost. *Sustainability, 14*(7), 3769. https://doi.org/10.3390/su14073769

Theodoulidis, B., Diaz, D., Crotto, F., & Rancati, E. (2017). Exploring corporate social responsibility and financial performance through stakeholder theory in the tourism industries. *Tourism Management, 62*, 173–188. https://doi.org/10.1016/j.tourman.2017.03.018

Ting, I. W. K., Azizan, N. A., Bhaskaran, R. K., & Sukumaran, S. K. (2019). Corporate social performance and firm performance: Comparative study among developed and emerging market firms. *Sustainability, 12*(1), 1–21. https://doi.org/10.3390/su12010026

White, H. (1980). A heteroskedasticity-consistent covariance matrix estimator and a direct test for heteroskedasticity. *Econometrica: Journal of the Econometric Society, 48*, 817–838. https://doi.org/10.2307/1912934

Whitelock, V. G. (2019). Multidimensional environmental social governance sustainability framework: Integration, using a purchasing, operations, and supply chain management context. *Sustainable Development, 27*(5), 923–931. https://doi.org/10.1002/sd.1951

Zahid, R. A., Khan, M. K., Anwar, W., & Maqsood, U. S. (2022). The role of audit quality in the ESG-corporate financial performance nexus: Empirical evidence from Western European companies. *Borsa Istanbul Review, 22*, S200–S212. https://doi.org/10.1016/j.bir.2022.08.011

Social Sustainability Management Practice in the Textile Industry

Budi Harsanto, Joval Ifghaniyafi Farras, and Yulistyne Kasumaningrum

Contents

1 Introduction

Recent studies show that an increasing number of consumers prefer to purchase from sustainable brands (Fatemi et al., 2023; Kim & Oh, 2020). However, the textile industry remains one of the most challenging sectors to make sustainable (Fadara & Wong, 2023; Harsanto et al., 2023). Among the triple bottom line of sustainability, social sustainability is of particular significance in ensuring equity and fairness among all stakeholders within the supply chain.

B. Harsanto (✉) · Y. Kasumaningrum
Faculty of Economics and Business, Universitas Padjadjaran, Bandung, Indonesia
e-mail: budi.harsanto@unpad.ac.id; yulistyne@unpad.ac.id

J. I. Farras
Digital Marketing Program, Universitas Padjadjaran, Bandung, Indonesia
e-mail: joval.ifghaniyafi@unpad.ac.id

S. S. Muthu (ed.), *Assessment of Social Sustainability Management in Various Sectors*, Environmental Footprints and Eco-design of Products and Processes,
https://doi.org/10.1007/978-3-031-81029-9_3

The concept of social sustainability is a broad one, encompassing the fulfilment of human needs, the conservation of the environment, advocacy for social equality, and the promotion of community well-being (Colantonio, 2010; De Fine Licht & Folland, 2019; Harsanto et al., 2024b). It is an integral part of sustainable development efforts and requires active engagement, fair opportunities, and the preservation of socio-cultural characteristics to foster a sustainable and thriving society (Boström, 2012; Eizenberg & Jabareen, 2017; Füsun Tatlıdil et al., 2009; Goel et al., 2020).

The journey of social sustainability in the textile industry has been extensive. Since the early stages of the Industrial Revolution, several significant achievements have been made, including the establishment of global labour standards and the implementation of corporate social responsibility (CSR) practices. Despite the progress made, challenges related to social sustainability continue. Issues such as poor working conditions and insufficient wages persist (Peake & Kenner, 2020). The global supply chains typical in the textile industry add complexity to the implementation of social sustainability management.

As the literature on social sustainability management in the textile industry evolves, several gaps remain. Current research focuses on the environmental aspect of sustainability or specific aspects of social sustainability management, such as barriers (Shaw et al., 2022) or social responsibility (Luque & Herrero-García, 2019). Some reviews have also been conducted in this area, but they again focus more on specific aspects, such as supply chain management (Köksal & Strähle, 2021). A comprehensive synthesis of various studies exploring social sustainability management practices in the textile industry is still limited. Thus, the aim of this paper is to uncover social sustainability management practices in the textile industry.

This chapter is structured into several sections. Following this section, the methodology used and an explanation of the study collection and analysis methods are discussed. The next section presents the results and discussion. The final chapter provides the conclusion.

2 Methodology

The systematic literature review approaches employed in the literature review process are referenced in a narrative approach to the methodology (Tranfield et al., 2003). More specifically, these reviews provide a structured and transparent way to synthesize and organize the available information on the topic (Tian et al., 2018). By defining explicit protocol instructions, systemic literature reviews enable researchers to perfectly search and evaluate studies relevant to research questions, thus leading to minimum bias and rigor in the research process (Tian et al., 2018). Moreover, systematic reviews differ from traditional narrative reviews in that they are replicable and scientific, which means they are involved in searching for published and unpublished primary studies and they generate an audit trail of decisions made during the review (Tranfield et al., 2003). This approach has been extensively employed in the field of business and management such as the studies from Harsanto et al. (2023, 2024a, b), Harsanto & Permana (2019), Sarasi et al., (2023), and Usman et al., (2022).

The data utilized in this research was gathered from the Scopus database. A literature search was conducted on April 2024 using the keywords ("social sustainability" or "social sustainability management") and ("textile" or "clothing" or "apparel"). The search yielded 305 documents, and we subsequently focused our analysis on articles, resulting in a total of 214 articles. Subsequently, the subject areas were filtered with a focus on business, management, accounting, social sciences, economics, econometrics, and finance, resulting in 155 articles. Following the application of the title/abstract filter, the remaining articles numbered 71. The focus was on articles that discussed social sustainability practices in the textile industry as the main keyword and were published in journals that were in quartile 1 and had an impact factor > 1. Articles that were not focused on social sustainability in the textile industry, were not published in quartile 1 journals, and had an impact factor < 1 were excluded from the analysis and synthesis, resulting in a final selection of 53 articles after a thorough full-text review. The flow of the article selection process based on the Preferred Reporting Items for Systematic Reviews and Meta-Analyses (PRISMA) approach can be seen in Fig. 1.

3 Results and Discussion

The total sample size is 53 articles from 9 different regions. The most common regions were from the United Kingdom (20%). The authors who are most cited by other researchers are Huq et al. (2016) from the United Kingdom, with 223 citations. Their research explores social sustainability practices within the specific circumstances in Bangladeshi clothing industry, emphasizing the dynamic relationship between global companies and their supply chain partners in developing countries. Furthermore, a research with a total of 214 citations was conducted by Raphaëlle Stewart and Monia Niero from Denmark. Their research discusses social sustainability practices in the traditional sense of fair labour practices or community development. However, the authors indirectly address the social implications of the circular economy (CE) by highlighting the importance of consumer behaviour and collaboration for the successful transition to a CE. Finally, research conducted by Wilhelm et al. from the Netherlands has been cited 179 times. Their research highlight that all buying firms in their sample actively extended sustainability practices to their tier 1 suppliers, often through long-term collaborative relationships. However, the approaches to managing sub-suppliers (tier 2 and beyond) varied significantly, ranging from direct involvement to complete delegation of responsibility. Authors present the most cited articles in Table 1.

This study examines research on social sustainability practices within the textile industry from 2009 to the present. The peak in publications on this topic occurred in 2020 and 2023, with nine articles each year, indicating a sustained high level of interest. In total, 53 articles published between 2009 and 2024 were included in the systematic literature review (Fig. 2).

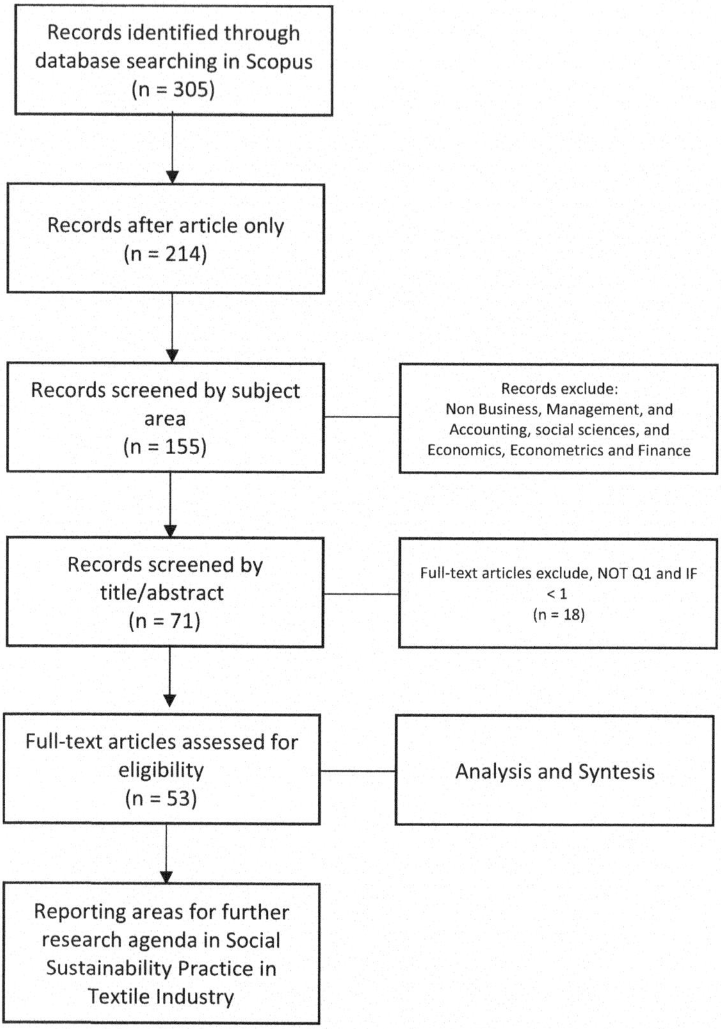

Fig. 1 A step-by-step procedure for choosing, analysing, and reporting articles. (Source: Adapted from Page et al., 2021)

All articles studied (53 articles) were from international journals (Q1) with impact factors in the range of 2.3–12.5, as illustrated in Table 2. The impact factor of 12.5 includes *Business Strategy and the Environment*, while the journal *Resources, Conservation and Recycling* has an impact factor of 11.2. Furthermore, the journal *Sustainable Development* has an impact factor of 9.9. The *International Journal of Production Economics* has an impact factor of 9.8. Journals that achieved an impact factor of 9.7 include the *Journal of Cleaner Production* and the journal *Corporate Social Responsibility and Environmental Management*; meanwhile, *Transportation Research Part E: Logistics and Transportation Review* has an impact factor of 8.3.

Table 1 Top 10 authors based on citations

Year	Authors	Cited by	Region
2016	Huq et al. (2016)	223	United Kingdom
2018	Stewart and Niero (2018)	214	Denmark
2016	Wilhelm et al. (2016)	179	Netherlands
2020	Cai and Choi (2020)	117	Hongkong
2016	Winter and Lasch (2016)	103	Germany
2021	Chowdhury and Quaddus (2021)	89	Australia
2021	Khan et al. (2021)	84	China
2019	Nayak et al. (2019)	81	Vietnam
2020	Taqi et al. (2020)	74	Bangladesh
2009	Hartlieb and Jones (2009)	74	United Kingdom

Source: Authors' own work

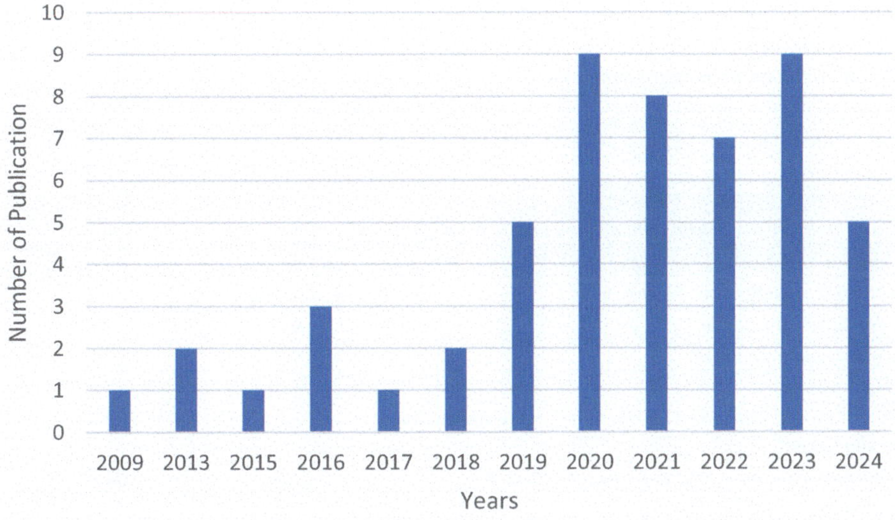

Fig. 2 Number of publication by year. (Source: Authors' own work)

4 Treemap Abstract

Analysis of the abstracts of the reviewed articles found that the three most frequently occurring terms were "sustainability" (188 times), "social" (186 times), and "supply" (142 times), representing 2.07, 2.04, and 1.56% of the total words, respectively, as shown in Fig. 3. This finding aligns with the focus of this research, which specifically explores social sustainability practices in the textile industry. The frequency of these words emphasizes the relevance and significance of this research in the context of social sustainability in the textile industry. Furthermore, the dominance of the word "sustainability" indicates that the concept has become a major

Table 2 Publication outlet

Journal	Impact factor	Rank
Business strategy and the environment	12.5	Q1
Resources, conservation and recycling	11.2	Q1
Sustainable development	9.9	Q1
International journal of production economics	9.8	Q1
Journal of cleaner production	9.7	Q1
Transportation research part E: Logistics and transportation review	8.3	Q1
Corporate social responsibility and environmental management	8.3	Q1
Journal of supply chain management	7.9	Q1
International journal of logistics management	7.2	Q1
International journal of operations and production management	7.1	Q1
International journal of production research	7	Q1
Operations management research	6.9	Q1
Journal of operations management	6.5	Q1
Journal of business ethics	5.9	Q1
International journal of physical distribution and logistics management	5.9	Q1
Corporate governance (Bingley)	5.5	Q1
Sustainability accounting, management and policy journal	5.2	Q1
Journal of industrial ecology	4.9	Q1
Business strategy and development	4.8	Q1
Management science	4.6	Q1
Management decision	4.1	Q1
TQM journal	3.8	Q1
International journal of productivity and performance management	3.6	Q1
Journal of business and industrial marketing	3.6	Q1
Sustainability (Switzerland)	3.3	Q1
Social responsibility journal	2.9	Q1
Clothing and textiles research journal	2.4	Q1
Journal of Asia business studies	2.3	Q1

Source: Authors' own work

concern in the industry, while the word "social" underlines the critical importance of understanding social dimension in achieving sustainability. Additionally, the occurrence of the word "supply" indicates that the industry's supply chain plays a crucial role in social sustainability practices.

Based on the articles reviewed, several social sustainability practices in the textile industry have been identified. We classified the identified practices into four themes, namely (1) safety working (SW), (2) labour condition and human rights (LHR), (3) society (S), and (4) product responsibility (PR), as shown in Table 3.

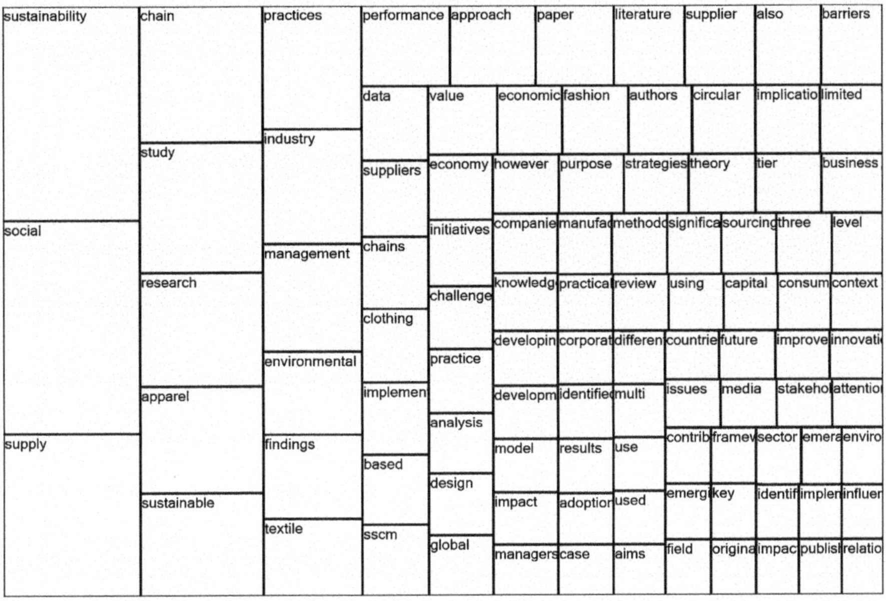

Fig. 3 Treemap abstract. (Source: Authors' own work)

The objective of this study is to conduct a comprehensive review of the existing research on social sustainability practices in the textile industry, which is still evolving, and to identify the practices that are currently being undertaken. The articles under review were all in the top quartile of Scopus and had an impact factor > 1. This demonstrates that the various studies conducted adhere to the highest standards of scientific rigour and have the potential to have a significant impact on research in the field of social sustainability, particularly in the context of the textile industry, which plays a pivotal role in the economy.

The papers analysed yielded several social sustainability practices in the textile industry. These were classified into four themes: (1) safety working (SW), (2) labour conditions and human rights (LHR), (3) society (S), and (4) product responsibility (PR).

5 Safety Working

Safety working is important for workers because it relates to safety during work in the factory. For example, programmes are very focused on ensuring employees feel safe during work, such as the implementation of factory safety initiatives, such as the agreement on Fire and Building Safety in Bangladesh and the Alliance for Worker Safety (Akbar & Ahsan, 2020; Susitha & Nanayakkara, 2023). The programme had an impact on improving working conditions and preventing accidents

Table 3 Social sustainability management practice

NO	Author	Safety working	Labour condition and human rights	Society	Product responsibility	Context	Impact
1	Zhang and Dong (2021)	X	X	X	√	China	Promote and expand various ways to reuse clothing, including selling refurbished second-hand items, offering swapping and rental services, and providing consultation on matching and styling old clothes
2	Pande and Adil (2022)	√	√	√	√	India	Sustainable manufacturing operations and employee development increased
3	Akbar and Ahsan (2020)	√	√	√	√	Bangladesh	Not only enhances the social performance of the firms but also boosts their competitive advantage in the global market
4	Chowdhury and Quaddus (2021)	√	√	√	√	Bangladesh	Improved market performance by reducing sustainability risks, enhancing social and environmental performance, and ensuring compliance with international standards

(continued)

Table 3 (continued)

NO	Author	Safety working	Labour condition and human rights	Society	Product responsibility	Context	Impact
5	Nath et al. (2023)	√	√	√	X	Bangladesh	Improved safety, reduced turnover, better productivity, but regulatory gaps remain
6	Caro et al. (n.d.)	√	√	√	X	Bangladesh	Reducing employee turnover, increasing workers' commitment, increase of factory productivity, reducing cost, and risk management
7	Najjar et al. (2024)	√	√	√	√	Global	Improved performance outcomes, including increase company profitability, supplier efficiency, and supplier innovativeness
8	Jung and Oh (2019)	√	X	√	√	China and South Korea	Enhance customer effectiveness on sustainability consumption behaviour
9	Susitha and Nanayakkara (2023)	√	√	√	X	Sri Lanka	Improvements in economic, social, and environmental performance metrics

(continued)

Table 3 (continued)

NO	Author	Safety working	Labour condition and human rights	Society	Product responsibility	Context	Impact
10	Esbeih et al. (2021)	√	X	√	√	Spain	Efficient waste management systems
11	Hamja et al. (2022)	√	√	X	X	Bangladesh	Reduced pain and discomfort among workers
12	Hartmann (2020)	√	X	X	X	Bangladesh	Increased accountability and improvements in working conditions as companies strive to meet higher standards and avoid negative publicity
13	Bubicz et al. (2021)	√	√	X	√	Global	Enhance labour conditions, human rights, and sustainability
14	García-Alaminos et al. (2024)	X	√	√	√	China and India	More ethical and responsible
15	Howard et al. (2022)	√	√	√	X	China and the United Kingdom	SMEs can not only improve their environmental and economic performance but also contribute to the Well-being of their employees and to the communities in which they do business
16	Cai and Choi (2020)	X	X	X	√ (sustainable design)	Developing country	Improve clothes design for a long-time use

(continued)

Table 3 (continued)

NO	Author	Safety working	Labour condition and human rights	Society	Product responsibility	Context	Impact
17	Hohn and Durach (2021)	√	X	√	√	Germany, United Kingdom, the Netherlands, India, Italy, and Bangladesh	New digital technologies like adaptive manufacturing might worsen existing social sustainability issues rather than improve them
18	Chowdhury et al. (2023)	√	√	√	X	Bangladesh	"Organizations can reduce costs by prioritizing sustainability barriers and choosing the most effective mitigation strategies."
19	Lis et al. (2020)	√	√	X	X	China, India, Iran, and Malaysia	More equitable and sustainable industry
20	Hrouga and Michel (2023)	√	√	X	√	France	More equitable and sustainable industry
21	Shaw et al. (2024)	X	√	√	X	India	Can move towards greater social sustainability, creating a supply chain that is characterized by fairness, safety, and respect for human rights
22	Hossain et al. (2022)	√	X	√	X	Bangladesh	Corporate sustainability is influenced by factors like company structure, its culture, social capital, and leadership approaches

(continued)

Table 3 (continued)

NO	Author	Safety working	Labour condition and human rights	Society	Product responsibility	Context	Impact
23	Zhang et al. (2023)	X	X	√	√	Bangladesh	A positive work environment, encouraging employee engagement and implementing quality management practices, to improve companies' social performance and contribution for sustainable future
24	Uddin et al. (2023)	√	√	X	X	Bangladesh	The commitment of the management team to the principles of social responsibility has an impact on the entire supply chain and contribute to a positive image of the organization, both locally and globally
25	Silvestre et al. (2020)	X	√	√	X	Global	Reduce the risk of corruption in the supply chain
26	Aytekin et al. (2023)	√	√	√	X	Türkiye	Improving employee welfare, strengthening relationships with stakeholders, and creating a positive impact on the local community

(continued)

Table 3 (continued)

NO	Author	Safety working	Labour condition and human rights	Society	Product responsibility	Context	Impact
27	Riemens et al. (2023)	X	√	X	√	Global	By improving traceability, companies can enhance transparency and accountability in their supply chains, which can help to promote social sustainability
28	Köksal and Strähle (2021)	X	X	√	X	Developing country	Social standards should be revised to include metrics that assess buying practices and capacity planning at factories, as well as the social performance of subcontractors
29	Camacho-Otero et al. (2020)	X	X	X	√	Colombia	Organizers can create more inclusive and engaging clothes swapping initiatives that contribute to social sustainability
30	Shaw et al. (2022)	√	X	√	X	India	Lead to improved working conditions, fair wages, and a positive impact on the communities involved in the supply chain

(continued)

Table 3 (continued)

NO	Author	Safety working	Labour condition and human rights	Society	Product responsibility	Context	Impact
31	Rovanto and Bask (2021)	X	√	√	√	Finland	Improved worker welfare, creation of positive outcomes for local communities, increased stakeholder engagement and support, and greater transparency in the communication of the company's social and environmental impacts
32	Ullah et al. (2024)	√	√	√	X	Pakistan	By fostering ORGI, firms can not only improve their environmental and economic performance but also their social impact and contribution on sustainable and equitable future
33	Galatti and Baruque-Ramos (2022)	√	√	X	X	Brazil	Emphasize that fair wages and benefits are essential for social sustainability and that companies should invest in the development of their workers and suppliers

(continued)

Table 3 (continued)

NO	Author	Safety working	Labour condition and human rights	Society	Product responsibility	Context	Impact
34	Sudusinghe and Seuring (2020)	√	√	√	X	Sri Lanka	Investing in social sustainability can lead to improved profitability, employee retention, and customer loyalty
35	Jajja et al. (2019)	√	X	√	√	Pakistan	Suppliers with a clan culture, which values collaboration and employee Well-being, are more likely to respond to normative pressures from industry peers and associations
36	Silva et al. (2023)	X	√	X	X	Brazil	The authors conclude that social sustainability in supply chains is not solely a matter of performance but also involves changing behaviours and values among supply chain members
37	Son et al. (2022)	X	X	X	√	The United States	Increase consumer awareness and interaction about sustainable products

(continued)

Table 3 (continued)

NO	Author	Safety working	Labour condition and human rights	Society	Product responsibility	Context	Impact
38	Nayak et al. (2019)	√	√	√	X	Vietnam	Growing awareness and commitment to social responsibility
39	Jayalath et al. (2024)	√	X	X	X	Sri Lanka	Apparel organizations can improve their social performance and enhance employee satisfaction
40	Boffelli et al. (2019)	X	X	√	X	Italy	Contribute to stakeholder engagement, collaboration, and social responsibility, all of which are essential for a socially sustainable future
41	Taqi et al. (2020)	√	X	X	X	Bangladesh	Take an active role in improving the social conditions for both their employees and the local communities
42	Khan et al. (2021)	X	X	√	√	China	Companies can create a more socially sustainable supply chain that benefits all stakeholders, including workers, communities, and the environment

(continued)

Table 3 (continued)

NO	Author	Safety working	Labour condition and human rights	Society	Product responsibility	Context	Impact
43	Arrigo (2020)	√	X	√	X	Italy	Benefit for companies and communities
44	Wilhelm et al. (2016)	√	X	X	X	The Netherlands	Improving the effectiveness of sustainability in the supply chain
45	Hansen and Schaltegger (2013)	X	√	X	√	Germany	Organic cotton production eliminates the health risks associated with pesticide exposure for farmers and workers
46	Köksal et al. (2018)	√	√	X	√	Vietnam	Improving social sustainability management
47	Winter and Lasch (2016)	√	√	X	X	Germany	Enhance reputation and consumer confidence
48	Hartlieb and Jones (2009)	√	√	√	√	The United Kingdom	Products obtained by consumers are more environmentally friendly
49	Ogilvy (2015)	√	√	√	√	Australia	Improved social Well-being, food security, and environmental protection, benefiting both present and future generations

(continued)

Table 3 (continued)

NO	Author	Safety working	Labour condition and human rights	Society	Product responsibility	Context	Impact
50	Shen et al. (2017)	√	√	√	√	China	Improve worker welfare, reduce negative environmental impacts, and create a more socially responsible and sustainable industry
51	Huq et al. (2016)	X	X	√	√	Bangladesh	Improved working conditions, regulatory compliance, and more ethical business practices
52	Stewart and Niero (2018)	X	X	√	X	China	Raising awareness, fostering collaboration, and promoting responsible consumption, all of which are essential for a socially sustainable future
53	Curwen et al. (2013)	√	√	X	√	Peru	Creating significant positive social impacts throughout EF's supply chain

Source: Authors' own work

in apparel manufacturing factories (Akbar & Ahsan, 2020). In addition, the following initiatives to ensure the safety and well-being of employees, such as providing safety training, protective equipment, and a safe working environment, are also expected to improve the safety and well-being of employees in the apparel industry (Chowdhury et al., 2023; Chowdhury & Quaddus, 2021). The same thing was also mentioned by Najjar et al. (2024), Nath et al. (2023), which state that ensuring workplace safety (fire, electrical, structural, and building safety), providing personal protective equipment (masks, earplugs, safety glasses, and needle guards), and promoting health, hygiene, and welfare (cleanliness, safe drinking water, ventilation, medical facilities, childcare, etc.) can provide a safe and healthy working environment for workers in clothing factories. Furthermore, Hamja et al. (2022) recommend integrating occupational health and safety (OHS) into lean initiatives, such as improving workplace design, reducing physical loads, and providing adequate safety equipment; with such integration comes improved workplace safety and ergonomics, which can reduce injuries and improve worker well-being.

The importance of safe working conditions should be emphasized with measures such as working hour controls, encouraging local labour, policies against harassment and abuse, workplace security controls, and worker mental health and care so as to improve worker safety and well-being and reduce the risk of workplace accidents and injuries (Galatti & Baruque-Ramos, 2022; Shen et al., 2017). Besides focusing only on workers, Winter & Lasch (2016) highlight the importance of the supplier's commitment to providing a safe and healthy working environment, including measures to prevent accidents, injuries, and exposure to hazardous substances. On the other hand, interesting things were mentioned by Hohn & Durach (2021). While automation is more socially sustainable than traditional sewing work, it requires more skilled labour, which leads to better working conditions and higher wages compared to unskilled manual labour.

In realizing safety working, companies began to focus on improving working conditions in their supply chains by providing training and development opportunities for workers and ensuring a safe and healthy working environment (Bubicz et al., 2021). Aytekin et al. (2023) further state that not only a safe and healthy working environment needs to be considered, but also an evaluation of the company's processes and operations needs to be carried out with regard to morality. So that in the end, the company will be ready to adopt and implement environmentally friendly safety working practices. Furthermore, Sudusinghe & Seuring (2020) emphasized the importance of providing nutritious food and concrete steps in safety working coupled with adequate safety equipment and sanitation facilities (Nayak et al., 2019). In order to strengthen safety working practices, Jajja et al. (2019) suggested implementing a Supplier Social Compliance Management System (SSCMS): Adoption and implementation of SSCMS, such as SA8000, ISO 26000, and Business Social Compliance Initiative (BSCI) codes.

6 Labour Condition and Human Rights

Labour conditions are one of the themes often considered in social sustainability management practices. For example, in India, training and development programmes are often implemented in the country, aimed at workers, managers, and suppliers on social sustainability issues that can raise awareness, improve understanding, and build capacity to implement sustainable practices (Shaw et al., 2022).

Furthermore, employee development practices support employee well-being and development, such as providing training and education opportunities, promoting diversity and inclusion, and ensuring fair compensation and benefits (Pande & Adil, 2022). The company also needs to provide appropriate and timely salary payments to employees as well as appropriate overtime pay and bonuses, as explained by (Nath et al., 2023; Nayak et al., 2019). In addition, companies need to offer health insurance and social protection for employees (Chowdhury & Quaddus, 2021; Nath et al., 2023). A similar programme is conducted in Australia, which aims to increase income and economic stability for farmers and rural communities, contributing to their overall well-being (Ogilvy, 2015).

Human rights are critical issues in the textile industry, particularly focusing on the prevention of child labour and forced labour. Previous studies, such as (Chowdhury & Quaddus, 2021; Nath et al., 2023), have highlighted this significant area of interest. In addition, the principles of equality and inclusiveness are utilized to ensure fair treatment of employees and equal opportunities for all (Nath et al., 2023; Sudusinghe & Seuring, 2020; Uddin et al., 2023). Women empowerment and equality promotion are issues of concern in the textile industry (Curwen et al., 2013; Nayak et al., 2019).

In order to protect human rights, organizations should encourage transparent communication and cooperation between management and employees (Galatti & Baruque-Ramos, 2022; Hamja et al., 2022), which aims to foster a positive and supportive workplace environment. These initiatives collectively illustrate the textile industry's commitment to upholding human rights and improving employee welfare so as to contribute to the social sustainability of the sector as a whole.

7 Society

The involvement of industry stakeholders may take the form of monetary supports for local community initiatives and actions on social sustainability practices in the textile industries (Lis et al., 2020). These could be financial aids for community activities, financing educational institutions, and engaging in philanthropic endeavours (Nath et al., 2023; Nayak et al., 2019; Sudusinghe & Seuring, 2020).

The textile industry needs to prioritize local workers such as considering the role of local communities to support the community (Chowdhury & Quaddus, 2021; Pande & Adil, 2022; Sudusinghe & Seuring, 2020). This is particularly important as the local workforce living in the vicinity of these companies is often overlooked. Furthermore, it is important to provide local communities with adequate sanitation facilities (Sudusinghe & Seuring, 2020).

8 Product Responsibility

Product responsibility is a social sustainability practice in the textile industry that needs to be considered so that products and production processes do not harm consumers and the environment. (Bubicz et al., 2021; Jung & Oh, 2019). Chemical and hazardous products can pose risk to consumer health if not considered. At the same time, organizations should strive to reduce the adverse environmental impact of their manufacturing operations. This is achieved through pollution reduction, recycling, water reuse (Silva et al., 2023), and recycle waste (Köksal et al., 2018).

On the other hand, the adoption of fair-trade labelling (Hartlieb & Jones, 2009) and fair-trade practices (Curwen et al., 2013) is an integral aspect of product responsibility; thus, supply chain transparency is also crucial to ensure that textile products are produced ethically and responsibly.

9 Conclusion

This paper explores social sustainability management practices within the textile industry. We identified publication trends, publication outlets, and top publications in this field. Four prominent practices are safety working, labour conditions and human rights, society, and product responsibility. Safety working practices, for instance, are implemented through work hour control, improving workplace design, ergonomics, and providing adequate safety enhancements. Labour conditions and human rights practices include supporting employee well-being and providing fair wages. Practices related to society involve engagement in supporting local community initiatives. Practices related to product responsibility focus on improving product safety, transparency, and sustainability.

This study has limitations. As a review based on published research, there may be practices not covered in the included articles. Future studies could include grey literature, such as industry reports and relevant websites, to complement these findings. Another limitation is that practices evolve quickly, so the reviewed publications might be outdated and not reflect the most current practices in the industry.

The implications of this analysis show that for companies, adopting social sustainability management practices can lead to positive impacts such as improved employee health and well-being, customer trust, a good public reputation, and compliance with international regulatory standards. For society, these practices also help create a more equitable and healthier society. For policymakers, these findings can encourage the crafting of regulations that support social sustainability practices for companies in the textile industry.

References

Akbar, S., & Ahsan, K. (2020). Investigation of the challenges of implementing social sustainability initiatives: A case study of the apparel industry. *Social Responsibility Journal, 17*(3), 343–366. https://doi.org/10.1108/SRJ-09-2019-0291

Arrigo, E. (2020). Global sourcing in fast fashion retailers: Sourcing locations and sustainability considerations. *Sustainability (Switzerland), 12*(2). https://doi.org/10.3390/su12020508

Aytekin, A., Okoth, B. O., Korucuk, S., Karamaşa, Ç., & Tirkolaee, E. B. (2023). A neutrosophic approach to evaluate the factors affecting performance and theory of sustainable supply chain management: Application to textile industry. *Management Decision, 61*(2), 506–529. https://doi.org/10.1108/MD-05-2022-0588

Boffelli, A., Dotti, S., Gaiardelli, P., Carissimi, G., & Resta, B. (2019). Corporate environmental management for the textile industry: Toward an empirical typology. *Sustainability (Switzerland), 11*(23). https://doi.org/10.3390/su11236688

Boström, M. (2012). A missing pillar? Challenges in theorizing and practicing social sustainability: Introduction to the special issue. *Sustainability: Science, Practice and Policy, 8*(1), 3–14. https://doi.org/10.1080/15487733.2012.11908080

Bubicz, M. E., Dias Barbosa-Póvoa, A. P. F., & Carvalho, A. (2021). Social sustainability management in the apparel supply chains. *Journal of Cleaner Production, 280*, 124214. https://doi.org/10.1016/j.jclepro.2020.124214

Cai, Y. J., & Choi, T. M. (2020). A United Nations' sustainable development goals perspective for sustainable textile and apparel supply chain management. *Transportation Research Part E: Logistics and Transportation Review, 141*, 102010. https://doi.org/10.1016/j.tre.2020.102010

Camacho-Otero, J., Pettersen, I. N., & Boks, C. (2020). Consumer engagement in the circular economy: Exploring clothes swapping in emerging economies from a social practice perspective. *Sustainable Development, 28*(1), 279–293. https://doi.org/10.1002/sd.2002

Caro, F., Lane, L., Sáez, A., & Cuenca, T. (n.d.). *Can brands claim ignorance? Unauthorized subcontracting in apparel supply chains*. https://ssrn.com/abstract=3621141

Chowdhury, M. M. H., & Quaddus, M. A. (2021). Supply chain sustainability practices and governance for mitigating sustainability risk and improving market performance: A dynamic capability perspective. *Journal of Cleaner Production, 278*, 123521. https://doi.org/10.1016/j.jclepro.2020.123521

Chowdhury, M. M. H., Rahman, S., Quaddus, M. A., & Shi, Y. (2023). Strategies to mitigate barriers to supply chain sustainability: An apparel manufacturing case study. *Journal of Business and Industrial Marketing, 38*(4), 869–885. https://doi.org/10.1108/JBIM-04-2021-0233

Colantonio, A. (2010). Urban social sustainability themes and assessment methods. *Proceedings of the Institution of Civil Engineers – Urban Design and Planning, 163*(2), 79–88. https://doi.org/10.1680/udap.2010.163.2.79

Curwen, L. G., Park, J., & Sarkar, A. K. (2013). Challenges and solutions of sustainable apparel product development: A case study of Eileen Fisher. *Clothing and Textiles Research Journal, 31*(1), 32–47. https://doi.org/10.1177/0887302X12472724

De Fine Licht, K., & Folland, A. (2019). Defining "social sustainability": Towards a sustainable solution to the conceptual confusion. *Etikk i Praksis – Nordic Journal of Applied Ethics, 2*, 21–39. https://doi.org/10.5324/eip.v13i2.2913

Eizenberg, E., & Jabareen, Y. (2017). Social sustainability: A new conceptual framework. *Sustainability, 9*(1), 68. https://doi.org/10.3390/su9010068

Esbeih, K. N., Molina-Moreno, V., Núñez-Cacho, P., & Silva-Santos, B. (2021). Transition to the circular economy in the fashion industry: The case of the inditex family business. *Sustainability (Switzerland), 13*(18). https://doi.org/10.3390/su131810202

Fadara, T. G., & Wong, K. Y. (2023). A decision support system for sustainable textile product assessment. *Textile Research Journal, 93*(9–10), 1971–1989. https://doi.org/10.1177/00405175221135167

Fatemi, H., Leijerholt, U., Rezvani, Z., & Schnittka, O. (2023). Consumer responses to sustainable product branding strategies: A literature review and future research agenda. *Baltic Journal of Management, 18*(4), 525–542. https://doi.org/10.1108/BJM-11-2022-0412

Füsun Tatlıdil, F., Boz, İ., & Tatlidil, H. (2009). Farmers' perception of sustainable agriculture and its determinants: A case study in Kahramanmaras province of Turkey. *Environment, Development and Sustainability, 11*(6), 1091–1106. https://doi.org/10.1007/s10668-008-9168-x

Galatti, L. G., & Baruque-Ramos, J. (2022). Circular economy indicators for measuring social innovation in the Brazilian textile and fashion industry. *Journal of Cleaner Production, 363*, 132485. https://doi.org/10.1016/j.jclepro.2022.132485

García-Alaminos, Á., Monsalve, F., & Zafrilla, J. (2024). Disentangling social impacts in global value chains through structural path analysis: Investigating forced labor in the cotton industry. *Journal of Industrial Ecology, 28*(2), 303–319. https://doi.org/10.1111/jiec.13359

Goel, A., Ganesh, L. S., & Kaur, A. (2020). Social sustainability considerations in construction project feasibility study: A stakeholder salience perspective. *Engineering, Construction and Architectural Management, 27*(7), 1429–1459. https://doi.org/10.1108/ECAM-06-2019-0319

Hamja, A., Hasle, P., & Hansen, D. (2022). Transfer mechanisms for lean implementation with OHS integration in the garment industry. *International Journal of Productivity and Performance Management, 71*(8), 3534–3555. https://doi.org/10.1108/IJPPM-06-2020-0314

Hansen, E. G., & Schaltegger, S. (2013). 100 per cent organic? A sustainable entrepreneurship perspective on the diffusion of organic clothing. *Corporate Governance (Bingley), 13*(5), 583–598. https://doi.org/10.1108/CG-06-2013-0074

Harsanto, B., & Permana, C. (2019). Understanding sustainability-oriented innovation (SOI) using network perspective in Asia Pacific and ASEAN: A systematic review. *JAS (Journal of ASEAN Studies), 7*(1), 1. https://doi.org/10.21512/jas.v7i1.5756

Harsanto, B., Primiana, I., Sarasi, V., & Satyakti, Y. (2023). Sustainability innovation in the textile industry: A systematic review. *Sustainability, 15*(2), 1549. https://doi.org/10.3390/su15021549

Harsanto, B., Farras, J. I., Firmansyah, E. A., Pradana, M., & Apriliadi, A. (2024a). Digital technology 4.0 on halal supply chain: A systematic review. *Logistics, 8*(1), 21. https://doi.org/10.3390/logistics8010021

Harsanto, B., Farras, J. I., & Indradi, D. (2024b). Sustainability innovation in the textile industry. In *Industry and innovation: Textile industry*. Springer Nature.

Hartlieb, S., & Jones, B. (2009). Humanising business through ethical labelling: Progress and paradoxes in the UK. *Journal of Business Ethics, 88*(3), 583–600. https://doi.org/10.1007/s10551-009-0125-x

Hartmann, J. (2020). Toward a more complete theory of sustainable supply chain management: The role of media attention. *Supply Chain Management, 26*(4), 532–547. https://doi.org/10.1108/SCM-01-2020-0043

Hohn, M. M., & Durach, C. F. (2021). Additive manufacturing in the apparel supply chain – Impact on supply chain governance and social sustainability. *International Journal of Operations and Production Management, 41*(7), 1035–1059. https://doi.org/10.1108/IJOPM-09-2020-0654

Hossain, M. B., Nassar, S., Rahman, M. U., Dunay, A., & Illés, C. B. (2022). Exploring the mediating role of knowledge management practices to corporate sustainability. *Journal of Cleaner Production, 374*, 133869. https://doi.org/10.1016/j.jclepro.2022.133869

Howard, M., Yan, X., Mustafee, N., Charnley, F., Böhm, S., & Pascucci, S. (2022). Going beyond waste reduction: Exploring tools and methods for circular economy adoption in small-medium enterprises. *Resources, Conservation and Recycling, 182*, 106345. https://doi.org/10.1016/j.resconrec.2022.106345

Hrouga, M., & Michel, S. (2023). Towards a new supply chain manager dashboard under circular economy constraints: A case study on France textile and clothing industry. *Business Strategy and the Environment, 32*(8), 6074–6093. https://doi.org/10.1002/bse.3473

Huq, F. A., Chowdhury, I. N., & Klassen, R. D. (2016). Social management capabilities of multinational buying firms and their emerging market suppliers: An exploratory study of the clothing industry. *Journal of Operations Management, 46*, 19–37. https://doi.org/10.1016/j.jom.2016.07.005

Jajja, M. S. S., Asif, M., Montabon, F. L., & Chatha, K. A. (2019). The influence of institutional pressures and organization culture on supplier social compliance management systems. *International Journal of Physical Distribution and Logistics Management, 49*(5), 552–574. https://doi.org/10.1108/IJPDLM-11-2017-0359

Jayalath, M. M., Perera, H. N., Seuring, S., & Thibbotuwawa, A. (2024). Social drivers affecting job design in apparel supply chains: Inferences from a discrete choice experiment. *Corporate Social Responsibility and Environmental Management, 31*, 3395. https://doi.org/10.1002/csr.2757

Jung, H. J., & Oh, K. W. (2019). Exploring the sustainability concepts regarding leather apparel in China and South Korea. *Sustainability (Switzerland), 11*(19). https://doi.org/10.3390/su11195389

Khan, S. A. R., Zkik, K., Belhadi, A., & Kamble, S. S. (2021). Evaluating barriers and solutions for social sustainability adoption in multi-tier supply chains. *International Journal of Production Research, 59*(11), 3378–3397. https://doi.org/10.1080/00207543.2021.1876271

Kim, Y., & Oh, K. W. (2020). Which consumer associations can build a sustainable fashion brand image? Evidence from fast fashion brands. *Sustainability, 12*(5), 1703. https://doi.org/10.3390/su12051703

Köksal, D., & Strähle, J. (2021). Social sustainability in fashion supply chains – Understanding social standard implementation failures in Vietnam and Indonesia using agency theory. *Sustainability (Switzerland), 13*(4), 1–35. https://doi.org/10.3390/su13042159

Köksal, D., Strähle, J., & Müller, M. (2018). Social sustainability in apparel supply chains-the role of the sourcing intermediary in a developing country. *Sustainability (Switzerland), 10*(4). https://doi.org/10.3390/su10041039

Lis, A., Sudolska, A., & Tomanek, M. (2020). Mapping research on sustainable supply-chain management. *Sustainability (Switzerland), 12*(10). https://doi.org/10.3390/SU12103987

Luque, A., & Herrero-García, N. (2019). How corporate social (ir)responsibility in the textile sector is defined, and its impact on ethical sustainability: An analysis of 133 concepts. *Corporate Social Responsibility and Environmental Management, 26*(6), 1285–1306. https://doi.org/10.1002/csr.1747

Najjar, M., Yasin, M. M., Bell, M. Z., & Bell, D. (2024). Investigating social sustainability practices in global supply networks: A fuzzy-set qualitative comparative analysis. *Journal of Cleaner Production, 447*, 141590. https://doi.org/10.1016/j.jclepro.2024.141590

Nath, S. D., Eweje, G., & Barua, S. (2023). Drivers and barriers for implementing social sustainability in supply chains: A qualitative investigation of a developing country's multi-tier suppliers. *International Journal of Logistics Management*. https://doi.org/10.1108/IJLM-03-2022-0121

Nayak, R., Akbari, M., & Maleki Far, S. (2019). Recent sustainable trends in Vietnam's fashion supply chain. *Journal of Cleaner Production, 225*, 291–303. https://doi.org/10.1016/j.jclepro.2019.03.239

Ogilvy, S. (2015). Developing the ecological balance sheet for agricultural sustainability. *Sustainability Accounting, Management and Policy Journal, 6*(2), 110–137. https://doi.org/10.1108/SAMPJ-07-2014-0040

Page, M. J., McKenzie, J. E., Bossuyt, P., Boutron, I., Hoffmann, T. C., Mulrow, C. D., Shamseer, L., Tetzlaff, J. M., Akl, E., Brennan, S. E., Chou, R., Glanville, J., Grimshaw, J. M., Hróbjartsson, A., Lalu, M. M., Li, T., Loder, E. W., Mayo-Wilson, E., McDonald, S., et al. (2021). The prisma 2020 statement: An updated guideline for reporting systematic reviews. *Medicina Fluminensis, 57*(4), 444–465. https://doi.org/10.21860/medflum2021_264903

Pande, B., & Adil, G. K. (2022). Assessment of the current state of sustainability in a manufacturing firm. *International Journal of Productivity and Performance Management, 71*(4), 1254–1276. https://doi.org/10.1108/IJPPM-04-2020-0151

Peake, K., & Kenner, J. (2020). 'Slaves to fashion' in Bangladesh and the EU: Promoting decent work? *European Labour Law Journal, 11*(2), 175. https://doi.org/10.1177/2031952520911064

Riemens, J., Lemieux, A. A., Lassagne, M., & Lamouri, S. (2023). Apprehending traceability implementation in support of sustainable value chains: A novel analysis framework for the fashion industry. *Journal of Cleaner Production, 414*, 137501. https://doi.org/10.1016/j.jclepro.2023.137501

Rovanto, I. K., & Bask, A. (2021). Systemic circular business model application at the company, supply chain and society levels – A view into circular economy native and adopter companies. *Business Strategy and the Environment, 30*(2), 1153–1173. https://doi.org/10.1002/bse.2677

Sarasi, V., Primiana, I., Harsanto, B., & Satyakti, Y. (2023). Sustainable supply chain of Indonesia's textile & apparel industry: Opportunities and challenges. *Research Journal of Textile and Apparel*.

Shaw, M., Majumdar, A., & Govindan, K. (2022). Barriers of social sustainability: An improved interpretive structural model of Indian textile and clothing supply chain. *Sustainable Development, 30*(6), 1616–1633. https://doi.org/10.1002/sd.2331

Shaw, M., Majumdar, A., & Govindan, K. (2024). How are the barriers of social sustainability perceived in a multi-tier supply chain? A case of textile and clothing industry. *Operations Management Research, 17*(1), 91–113. https://doi.org/10.1007/s12063-023-00406-8

Shen, B., Li, Q., Dong, C., & Perry, P. (2017). Sustainability issues in textile and apparel supply chains. *Sustainability (Switzerland), 9*(9). https://doi.org/10.3390/su9091592

Silva, M. E., Mikosz, K. S. C., & Dias, G. P. (2023). A micro-level perspective for sustainability implementation in supply management: When operating context fosters opportunistic behaviours. *Journal of Cleaner Production, 414*, 137764. https://doi.org/10.1016/j.jclepro.2023.137764

Silvestre, B. S., Viana, F. L. E., & de Sousa Monteiro, M. (2020). Supply chain corruption practices circumventing sustainability standards: Wolves in sheep's clothing. *International Journal of Operations and Production Management, 40*(12), 1873–1907. https://doi.org/10.1108/IJOPM-06-2019-0454

Son, J., Nam, C., & Diddi, S. (2022). Emotion or information: What makes consumers communicate about sustainable apparel products on social media? *Sustainability (Switzerland), 14*(5). https://doi.org/10.3390/su14052849

Stewart, R., & Niero, M. (2018). Circular economy in corporate sustainability strategies: A review of corporate sustainability reports in the fast-moving consumer goods sector. *Business Strategy and the Environment, 27*(7), 1005–1022. https://doi.org/10.1002/bse.2048

Sudusinghe, J. I., & Seuring, S. (2020). Social sustainability empowering the economic sustainability in the global apparel supply chain. *Sustainability (Switzerland), 12*(7). https://doi.org/10.3390/su12072595

Susitha, E., & Nanayakkara, M. (2023). Impact of green supply chain management practices on the triple bottom line: A study on apparel manufacturers of Sri Lanka. *Journal of Asia Business Studies, 17*(6), 1228–1249. https://doi.org/10.1108/JABS-05-2022-0180

Taqi, H. M. M., Ahmed, H. N., Paul, S., Garshasbi, M., Ali, S. M., Kabir, G., & Paul, S. K. (2020). Strategies to manage the impacts of the COVID-19 pandemic in the supply chain: Implications for improving economic and social sustainability. *Sustainability (Switzerland), 12*(22), 1–25. https://doi.org/10.3390/su12229483

Tian, M., Deng, P., Zhang, Y., & Salmador, M. P. (2018). How does culture influence innovation? A systematic literature review. *Management Decision, 56*(5), 1088–1107. https://doi.org/10.1108/md-05-2017-0462

Tranfield, D., Denyer, D., & Smart, P. (2003). Towards a methodology for developing evidence-informed management knowledge by means of systematic review. *British Journal of Management, 14*(3), 207–222. https://doi.org/10.1111/1467-8551.00375

Uddin, M. H., Razzak, M. R., & Rahman, A. A. (2023). Sustainable supply chain management practices, dynamic capabilities and competitive advantage: Evidence from Bangladesh ready-made garments industry. *Business Strategy and Development, 6*(2), 176–188. https://doi.org/10.1002/bsd2.232

Ullah, S., Ahmad, T., Kukreti, M., Sami, A., & Shaukat, M. R. (2024). How organizational readiness for green innovation, green innovation performance and knowledge integration affects sustainability performance of exporting firms. *Journal of Asia Business Studies, 18*(2), 519–537. https://doi.org/10.1108/JABS-02-2023-0056

Usman, A., Azis, Y., Harsanto, B., & Azis, A. M. (2022). Airport service quality dimension and measurement: A systematic literature review and future research agenda. *International Journal of Quality and Reliability Management, 39*(10), 2302–2322. https://doi.org/10.1108/IJQRM-07-2021-0198

Wilhelm, M., Blome, C., Wieck, E., & Xiao, C. Y. (2016). Implementing sustainability in multi-tier supply chains: Strategies and contingencies in managing sub-suppliers. *International Journal of Production Economics, 182*, 196–212. https://doi.org/10.1016/j.ijpe.2016.08.006

Winter, S., & Lasch, R. (2016). Environmental and social criteria in supplier evaluation – Lessons from the fashion and apparel industry. *Journal of Cleaner Production, 139*, 175–190. https://doi.org/10.1016/j.jclepro.2016.07.201

Zhang, X., & Dong, F. (2021). How virtual social capital affects behavioral intention of sustainable clothing consumption pattern in developing economies? A case study of China. *Resources, Conservation and Recycling, 170*, 105616. https://doi.org/10.1016/j.resconrec.2021.105616

Zhang, S., Rew, D., Jung, J., Wu, S., & Baldo, C. (2023). Impact of organizational citizenship behavior on corporate sustainability through the mediation of TQM: Focus on the textile industry in Bangladesh. *TQM Journal*. https://doi.org/10.1108/TQM-02-2023-0056

Sufficiency Economy Philosophy in Fostering Agricultural Enterprise Well-being: A Case Study from Thailand

Ruethai Onbhuddha and Seiichi Ogata

Contents

R. Onbhuddha · S. Ogata (✉)
Graduate School of Energy Science, Kyoto University, Kyoto, Japan
e-mail: ruethai.onbhuddha.47i@st.kyoto-u.ac.jp; ogata.seiichi.5x@kyoto-u.ac.jp

S. S. Muthu (ed.), *Assessment of Social Sustainability Management in Various Sectors*, Environmental Footprints and Eco-design of Products and Processes,
https://doi.org/10.1007/978-3-031-81029-9_4

1 Introduction

Globalization, amid many crises, presents quick attacks on business and everyday personal life. Organizations and individuals must adjust to build their survival ability in the long term. In the face of a dramatically evolving global landscape, businesses must develop resilience in various dimensions, particularly in the wake of significant disruptions, such as the COVID-19 pandemic (Nguyen et al., 2023). Studies on corporate responsibility have focused on the social responsibilities of "the businessman" since 1953. Research has developed continuously since then to include corporate social responsibility (CSR), stakeholder theory, corporate sustainability, and alternative economics (Chang et al., 2017). Therefore, a sustainable business model requires balancing stakeholders, self-sufficiency, environmental impact, and time frame applications to create shared value (Norris, 2024).

The majority of businesses in Thailand are small- and medium-sized enterprises (SMEs), whereas microenterprises are prominent across all firm types (OECD, 2022). SMEs' adaptation to positive impacts can support sustainable growth (Gomes & Pinho, 2023; Jiang et al., 2023). Over a decade in Thailand, community enterprises have increased (Thailand enterprise amounts from 201 to 2022, 2022). This activity aims to support business growth at a grassroots level. Community enterprises benefit from promoting local income expansion and stronger collaborations (Kwangsawad & Jattamart, 2022), especially for agricultural businesses, which are essential to the food supply chain.

Based on rational thinking integration with consequential effects, sufficiency economy philosophy (SEP) is a principle that can be applied to Thailand's economic development (MFA, 2017). This philosophy presents an alternative way of thinking that should be integrated within organizational management, mainly focusing on the middle path context. These applications can be applied to construct a sustainable culture (Ketprapakorn & Kantabutra, 2019a). A leader's characteristics that align with SEP are related to positive company performance (Kantabutra & Siebenhuner, 2011). Moreover, employees are also important stakeholders in an organization, as their behaviors can affect corporate sustainability performance (Ruiz-Pérez et al., 2021). However, studies on employees' SEP practices within enterprises are limited. Therefore, this chapter evaluates critical factors based on SEP in agricultural enterprise practices. Specifically, we focus on the impact of employees, of which more than 30% of the total Thai workforce is employed in the agricultural sector (Bank of Thailand, 2024a). Evidently, agricultural enterprises are essential in supporting the bio-circular green model policy in Thailand (BCG's Research Community, 2018).

This study examined the significant relationship between SEP practices and the sustainability outcome of agricultural SMEs that have integrated SEP thinking within their business processes. This empirical research was conducted using a qualitative case study investigative approach. The remainder of this chapter is organized as follows. Section 2 provides an extensive review of the literature to build a conceptual understanding of SEP and the well-being of organizations and SMEs in

Thailand. Section 3 presents the research methodology used in this study. Section 4 includes the results and findings. Section 5 provides the discussion. Section 6 presents the research conclusions and study limitations.

2 Literature Review

2.1 Sufficiency Economy Philosophy (SEP)

The concept of SEP was initiated by Bhumibol Adulyadej, His Majesty the King, in the 1970s. However, SEP in Thailand gained widespread attention after the Asian financial crisis in 1997 (Song, 2020). SEP is a way of thinking that supports decision-making in sustainable development (Mongsawad & Thongpakde, 2016). Its applications can be used on multiple levels, including individual, community, organizational, and national scales. At the macro level, the national economy and social development plan from the ninth to the current 13th plan have integrated the SEP concept as a foundation of national development. One strategy of the National 13th Plan is to "drive the rehabilitating of natural resources and increasing resource efficiency based on the Sufficiency Economy Philosophy" (NESDC, 2022). This strategy illustrates the critical, holistic thinking of SEP related to Thailand's development. Regarding SEP guidance for industries, it was launched to promote the balance and sustainable development of organizations based on sufficient economic principles (TISI, 2013). Its voluntary guidelines indicate the industry's good practices and company activity.

The establishment of the SEP process (Bergsteiner & Piboolsravut, 2016) begins with the fundamental of virtues combined with knowledge to create an appropriate attitude. Subsequently, it guides wise decision-making and actions. Each SEP element embodies moderation, reasonableness, and self-immunity, reflecting good business practices, as shown in Table 1. The SEP framework illustrates the path toward sustainable outcomes, as shown in Fig. 1.

The SEP is appropriate for embedding and supporting everyone's way of thinking within an organization. As such, SEP supports organizational development in terms of long-term sustainability and self-resilience (Akkarangkul, 2019). A necessary SEP foundation, i.e., knowledge and virtue, are essential principles to consider for long-term enterprise operations (Songsom, 2014). Moreover, SEP organizations prioritize concern for the consequential impact on stakeholders instead of focusing merely on the shareholder (Kantabutra & Siebenhuner, 2011). This concept aligns with stakeholder theory (Freeman, 2010). As such, SEP characteristics can address sustainability on a large scale. Previous research indicates that SEP practices tend to achieve strong financial performance and resilience (Kantabutra, 2019; Kantabutra & Siebenhuner, 2011; Ketprapakorn & Kantabutra, 2019b; Korphaibool et al., 2021). Nevertheless, studies on the outcomes of well-being based on SEP in agricultural enterprises are limited. Therefore, this study aims to shed light on the

Table 1 SEP principles in business practices

Principle	Practice
Moderation	Manufacture or provide services at a level commensurate with the inherent capabilities of the business Refrain from the extremes of overconsumption or underconsumption Promote the efficient and adequate consumption of resources Focus on the consumption of local and available resources Establish appropriate pricing strategies for products and services Provide decision-making support for long-term business Establish good business partners and positive relationships with stakeholders Avoid the extreme in one direction: focus on economic coexistence with society and the environment
Reasonableness	Make decisions and actions based on reasonably based behavior Make decisions based on self-awareness of the business and market Business development by growth from within Use analytical capability and foresight to assess the consequential impact of business activities on business performance and stakeholders Emphasize rationale utilization in decision-making and actions
Self-immunity	Track business performance across various aspects such as economics, society, and the environment Continue to observe internal and external change to conduct risk management Continue monitoring opportunities, threats, and risks in business activities Diversify the product and service variety Make decisions with prudence, such as in material management and sourcing
Virtue	Adopt virtuous behavior that includes honesty, perseverance, sharing, loving, self-reliance, integrity, altruism, and public-mindedness within business processes Apply virtuous behavior in interconnecting with their stakeholders, such as CSR Integrate good corporate governance within company management processes Comply with relevant laws and regulations
Knowledge	Adopt the new knowledge that aligns with local geographic, social, and cultural contexts in business activities Provide the learning training for both operating process and individual development Develop further from existing knowledge in the organization Focus on innovation and research and development (R&D)

Source: Jitsuchon (2019), Kantabutra and Punnakitikashem (2020), Korphaibool et al. (2021), Petison and Kantabutra (2022), Song (2020)

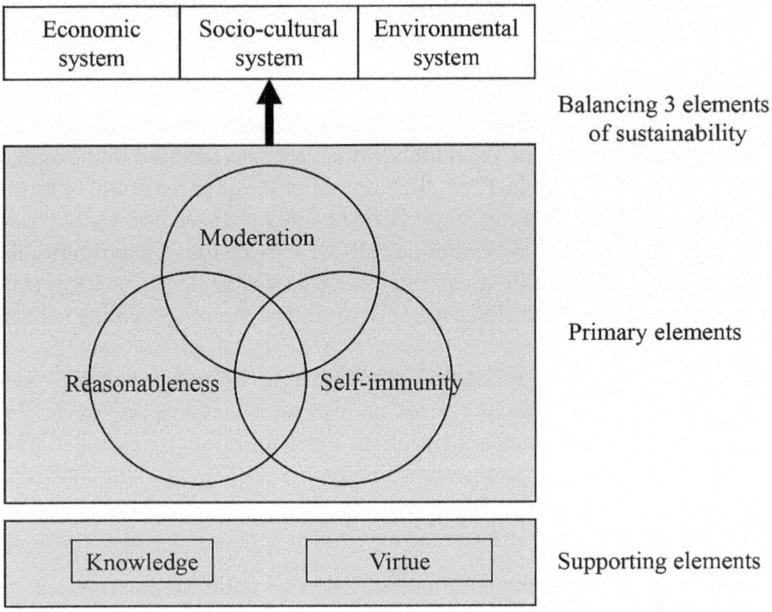

Fig. 1 This framework is adopted from Song, 2020

impact of the well-being dimension to illustrate the positive effects of balancing long-term performance.

2.2 *Well-being in Business*

The interconnection between well-being and business across various dimensions within an enterprise can increase humanity and ecological quality instead of focusing only on economic growth (Fioramonti et al., 2022). For instance, well-being in an organization relates to income, work quality, health, knowledge, satisfaction, social relativity, and work–life balance, which positively impact corporate performance (Alkhodary, 2023). Notably, the importance of employee well-being (Isham et al., 2021) indicates that worker well-being positively affects productivity. In contrast, focusing solely on productivity growth can have a negative effect on worker well-being. At the same time, (OECD, 2023) declared that employee tension and work period overload were associated with negative effects on companies. However, previous research has found that the level of employee satisfaction with accomplishments is related to a company's financial flow performance (Uchida & Kino, 2021). These findings indicate that the quality of work–life balance and organizational performance are inevitably related.

2.3 *Micro, Small, and Medium Enterprises (MSMEs) in Thailand*

Thailand is an agricultural country and a leader in agricultural production exports. As of 2021, more than 31% of Thailand's workforce was situated in the agricultural industry (FAO, 2023). From 2012 to 2023, the number of agricultural enterprises in Thailand increased, in which the majority of businesses were microscale, as shown in Fig. 2. After the COVID-19 pandemic, MSMEs in Thailand's agricultural sector expanded, causing the employment rate to increase by 15% in 2021 (OSMEP, 2022). This suggests that MSMEs in the agricultural sector are essential to Thailand's structure.

Based on the above, past research is limited to the links between SEP practices and outcomes in agricultural enterprises, an essential business sector in Thailand. More than 33% of Thai workers are within the agricultural sector (Bank of Thailand, 2024b). However, there is a gap in the research on well-being outcomes regarding SEP enterprise characteristics. Therefore, this study developed the research questions as follows:

RQ1: What are the main characteristics of SEP in agricultural enterprises?
RQ2: What are the primary outcomes of SEP in agricultural enterprises?

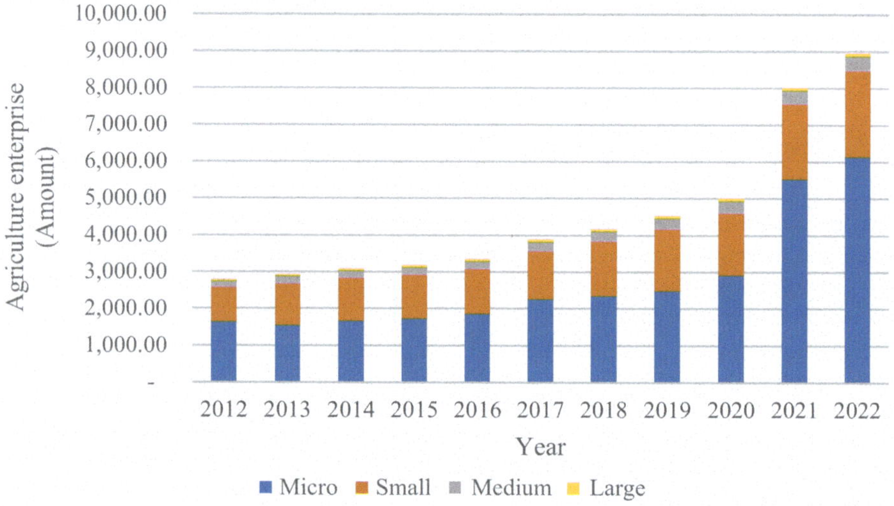

Fig. 2 Number of agricultural enterprises in Thailand from 2012 to 2023. Note: employees in micro businesses ≤5 persons, small businesses 6–50 persons, medium businesses 51–200 persons, large businesses ≥201 persons. Source: The Office of Small and Medium Enterprises Promotion (OSMEP), Thailand (Thailand enterprise amounts from 2021 to 2022″ 2022)

3 Research Methodology

3.1 Qualitative Approach Using Case Studies

This study adopted qualitative research based on a case study method to examine experiential data from the target group (Harsanto et al., 2022; Kantabutra, 2019). The case study approach aimed to elicit an in-depth understanding of the relationship between SEP and community enterprise management.

3.2 Case Study Profile

The case studies conducted in this study focus on SMEs that have adopted SEP in business management, including community and medium enterprises. These three case studies were conducted with agricultural businesses located in the northern part of Thailand. We selected enterprises from the north region that face high environmental impacts, especially PM 2.5 (Supasri et al., 2023). Figure 3 presents the geography of the community enterprises.

3.2.1 Enterprise A

Enterprise A is the Ban Ku Daeng community enterprise. This community enterprise aims to produce consumable products from local raw material and food. It was established in 2005 with the capital of THB 1.5 million. Their first product was detergent, offered at a lower price than conventional products to reduce household expenses. Since 2015, they have developed products from local rice, starting with germinated brown rice. In later years, powdered rice and porridge were respectively produced. The product processing area is situated as a common area within the community. The enterprise currently gathers 170 members from the same community. They hold a biannual meeting twice a year for formal enterprise communication. The enterprise allocates profit to six groups: 35% for the reserve fund, 20% for operators, 15% for member welfare, 10% for scholarships, 10% for community development, and 10% for enterprise research and development (R&D).

3.2.2 Enterprise B

Enterprise B, called Farm Hug community enterprise, produces nonchemical-based agricultural products. This enterprise was founded in 2022 after the COVID-19 crisis in Thailand. It was first registered with a capital of THB 0.01 million and comprised nine members initially. Products from this community consist of local vegetables, livestock products, fertilizer, and other farm by-products. This

Fig. 3 Geography of the case studies, Chiang Mai and Chiang Rai provinces, Thailand

enterprise applies biochar processing and sharing the kiln among community members. Furthermore, the learning center introduced zero-waste farm management to members and the public. Their meeting is set for twice a year for formal communication within the enterprise.

3.2.3 Enterprise C

Enterprise C, "Rai Ruen Rom," was established in 2014 and introduced a sustainable lifestyle with SEP perspectives in its business management (Gray, 2022). The owner started this project based on her passion for volunteer work while moving forward in joining the project named "1 Rai 100,000 baht," initiated by the Thai Chamber of Commerce and Bank for Agriculture and Agricultural Cooperatives (BAAC). The project was motivated by the SEP framework, and their business provides agricultural products, tourism, and services (Tipmongkol & Pooripakdee, 2022). Their business practices include plantations, food processing, dining, farm stays, and workshops in Chiang Rai province. Rai Ruen Rom started its business with planting. Two years later, the enterprise established a café and restaurant and sourced their own products to serve in the restaurant. The agricultural section includes plant seedlings, nurseries, plantations, maintenance, and harvesting. They produce soil improvements and fertilizer on their own, based on the materials obtained from residue within the area such as manure and agricultural waste.

In 2016, they built a farm stay and started generating renewable energy from photovoltaics with battery storage to replace fossil fuel consumption. Next, the farm invested in biogas from food waste and manure to expand its renewable energy capacity and to become more self-reliant. The solar house was constructed and provided as one of the farm stays, called a "cleaner energy farmhouse" until 2019. During COVID-19, they established a processing factory to expand their organic products to a broader range of consumers. The most recent service offering is the Roasty by Rai Ruen Rom, a restaurant with locations in Chiang Rai and Bangkok. Its newest business venture aims to serve its agricultural products within urban settings. Their market covers business to business (B to B) and business to customer (B to C). Currently, Rai Ruen Rom has a total 80 employees both on a monthly and daily basis. Every year, apprentices are open to train and learn more about the organic processes of the actual business.

Enterprises A and B are registered as community enterprises, while enterprise C is defined as a medium-sized enterprise. We selected these three enterprises because they all have demonstrated the application of the SEP principle in their activities. Therefore, this chapter focuses on analyzing the common characteristics of both community and medium-sized enterprises that integrate SEP into its business processes.

3.3 Triangulation Approach

Following Bekhet and Zauszniewski (2012) and Sepasgozar and Davis (2018), we adopted a triangulation approach to measure the qualitative analysis and compare findings from various data sources. Data was collected from primary and secondary sources through the triangulation method comprising (1) in-depth interviews, (2) field observations, and (3) secondary data.

3.3.1 Primary Source Data

The data collection process included the in-depth interview method with SEP community enterprises to determine the practices and concepts of the organization (Aschemann-Witzel et al., 2023; Rasiah & Ren, 2023). An in-depth survey of the case studies was also used to comprehensively explain the quality-based research (Carè et al., 2023; Oyelami & Olivier, 2015). In-depth interviews were approximately between 20 and 45 min per person and were digitally recorded. In-depth interviews took place at the case study site in the public area last September and

November 2023 and January 2024. Table 2 shows the number of interviewees and their position within the enterprise for each case study. The examples of questions for the focus group are shown in Table 3, and the question structure is presented in Appendix A.

After conducting in-depth interviews, we surveyed the agricultural and processing areas to enhance our understanding of the operational procedures within the organizational framework. Moreover, site visit observations were approximately 1 h and 30 min long, at which time we conducted a focus group. Meanwhile, we observed the activity and short interview of the value chain of Enterprise C for 5 hours to observe their process and staff's attitude.

3.3.2 Enterprise Observations

We selected the observation scheme to gain further insights into the enterprises' business process and staff's perspectives. The researcher considered the reflection based on short interviews and observation for a clear understanding of the corporate context (De Temmerman et al., 2023). The monitoring structure followed the SEP practices adopted from Table 1 and previous research. Table 4 shows the observation measurements.

3.3.3 Secondary Data Source

This study also collected data from annual enterprise activity reports, published reports, media, and information from the official website to support data in analyzing the process (Silva et al., 2021). The secondary data was obtained in the Thai language.

3.4 Data Analysis

Following the data collection process, we thoroughly analyzed the gathered data, focusing on grouping the management themes derived from the case study information. These management themes were systematically classified in accordance with

Table 2 Enterprise members who participated in in-depth interviews

Position	Enterprise A	Enterprise B	Enterprise C
Leader	1	1	1
Accountant	1		
Member		1	3
Manager			2
Total	2	2	6

Table 3 In-depth interview question example

No.	Characteristic	Business practices	Question
1.	Knowledge sharing	Knowledge sharing with other stakeholders is necessary to continue developing innovation and benefit customers and society (Kantabutra & Punnakitikashem, 2020).	How do you motivate partners to collaborate with the enterprise? How do you develop and innovate products?
2.	Moderation investment	The investment decisions are based on the deep understanding of business and customer and support the risk management (Petison & Kantabutra, 2022)	How do you decide on new investments?

Table 4 Enterprises' observation principle and measurement

Principle	Measurement
Moderation	Does the company have a production plan based on capacity and demand? Does the company focus on energy and material efficiency? Does the company focus on long-term business? What are the merits of creating a good partner? Does the company consider local resources?
Reasonableness	Is the company concerned about the consequential impact of energy and material consumption? Does the company focus on long-term mutual benefits of the enterprise and its stakeholders?
Self-immunity	Is the company concerned with internal and external risks? Does the company develop diversified products? Does the company depend on diversified resources? Does the company's financial plan refrain from over-debt?
Knowledge	Does the company consider investing in people development? Does the company focus on research and development? Does the company analyze in-house information and implementation to understand business?
Virtue	Does the company support teamwork in the working process? Does the company encourage its members to persevere in their work? Does the company apply good governance within the corporation?

the specifications outlined in Table 4. Next, we summarized the identified case study management themes alongside the SEP approach, aiming to draw conclusive insights into the primary characteristics emerging from the findings. We further analyzed the output and outcome of SEP enterprises. The research framework is presented in Fig. 4.

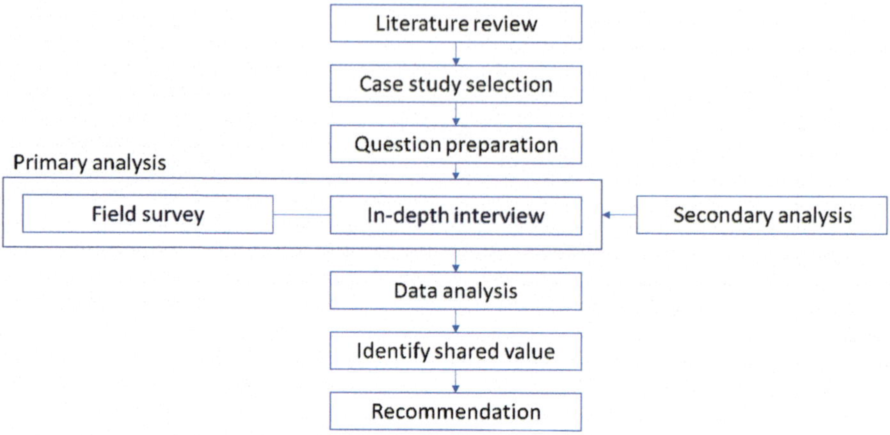

Fig. 4 Research framework

Table 5 SWOT analysis of the case studies

Strengths	Weaknesses
Immunity from debt prevention Quality control and quality certification Eco-friendly products Resource localization and self-sufficiency Resource efficiency consumption Partnerships Strong networking Better family quality	The worker limitation Late return on financial performance
Opportunities	**Threats**
Human development Net-zero business practices	Community attitudes changing Market fluctuation

4 Results

The findings from the triangulation technique demonstrate that SEP was related to focused enterprise practices. The research observations identified the strengths and weaknesses of the enterprises, as detailed in Table 5. Identifying these strengths and weaknesses served as an indicator of organizational self-awareness, providing insights that can be instrumental in recognizing areas for further improvement.

The results in Table 5 illustrate both positive and negative aspects regarding the enterprises based on the case studies. The positive elements were constructed from their understanding of their ability and potential. Additionally, they created value with their stakeholders. In contrast, the enterprises must cope with negative aspects from the dynamic world—the significant characteristics of SEP agricultural enterprises and outcomes are illustrated in the following section.

4.1 Key Success Characteristics in Agricultural Enterprises

The findings from the interviews and observations indicate that the three enterprises' practices align with SEP thinking. At the same time, SEP inputs reflect outcomes that increase the quality of life in society. The following section presents the main characteristics of the community enterprises that apply SEP while reflecting the positive outcomes to support economic, social, and environmental sustainability. The selected case studies' key success characteristics were willingness to adapt, moderation behavior, and sharing. The following discussion details the robust practices of the three enterprises.

4.1.1 Willingness to Adapt

Adaptation is the foundation of self-awareness. The enterprises improved to become more secure, which was the common practice across all case studies. However, this adaptation is also based on the organization's self-awareness in business, as illustrated in the following interview excerpts.

> "Before starting a new production, I explore the available raw materials and problems in a community to set the target product. After that, I will trial and error in the product development stage."
>
> "The learning local resource beneficial from local experts to select an appropriate herbal and medical plant to utilize in new products." (Enterprise A)

Resource and energy security are essential to an enterprise. It reflects self-immunity in business practices if an enterprise is open minded to installing renewable energy for upgrading their security in energy prices. In term of resources, the enterprises tried to source their resources from local areas as prioritizing raw material to ensure sufficient input.

> "To protect against high energy expenses in agricultural farming, we integrated renewable energy such as solar pumping and solar cells in the chicken farm to enhance the immune resilience of the farm." (Enterprise B)
>
> "We installed solar and biogas within our enterprise because we consider the long-term impact on our stakeholders. So, we must strengthen our immunity to fossil fuel dependency. We can consume our self-energy and reduce the import energy prices." (Enterprise C)

Integrating feedback from stakeholders into decision-making is a strength of every organization. Successful organizations listen to their customers and are willing to produce new products to solve customers' problems. However, during new product development, knowledge improvement within an organization is part of the process. Thus, a Plan-Do-Check-Act (PDCA) management strategy must be integrated into new product development.

> "There are many elders in our community. So, they need to buy new products which are easier to eat and more convenient. Therefore, I tried to develop rice powder for drinking." (Enterprise A)

> "During new product development, I continue trial and error to make good quality products before selling." (Enterprise A)
>
> "We keep tracking the farm performance during every cultivation; both input and output data are observed, and we apply the results for planning and improvement." (Enterprise C)

The willingness to adapt behavior was illustrated in all three case studies. These enterprises are challenged to develop new products, improve processes, and provide support based on data from monitoring and backup knowledge. For example, the charcoal project in Enterprise B and compost fertilizer project in Enterprise C involve monitoring data and continually improving systems and processes to be the most up to date with their utilization.

The significant behavior shown in the case studies reflects their willingness to adapt their processes. This aligns with the holistic reasonableness of SEP thinking. As such, the findings of this study confirm (Methamontri et al., 2022) that "perceived usefulness" is related to the willingness to install new energy systems such as agrivoltaic energy. This emphasizes that an enterprise pursuing value for long-term impact on stakeholders can alter its processes, as shown specifically in Enterprise C. This result is also supported by the research on environmental transition (Lennartz et al., 2022), where adaptation and value creation are consequential to new resource allocation in the value chain and the willingness to adapt, particularly regarding behavioral change resulting from reduced individual resource use (Hayles & Dean, 2015). Moreover, the findings reinforce that self-immunity in their system framework drives long-term performance. Therefore, adaptation is essential to business transformations toward a sustainable paradigm. However, the continual monitoring and improving with a reasonableness mindset are significant in business adaptation in adjusting to the most appropriate company process.

4.1.2 Operating Organizations with Moderation

The findings of the case study illustrates that the enterprises engaged in ongoing scrutiny of the accounting status within their organizations and encouragement of all members to monitor family expenditures. This aims to empower effective management of family finances, ensuring balanced income and expenditure. This strategy involves safeguarding investments solely through internal funds while refraining from seeking external financing. This approach is beneficial to protecting an enterprise from over-excessive debt. Strategic production planning is crucial for aligning with production capacity and customer demand. This approach serves to uphold stringent quality control measures and acts as a safeguard against overproduction.

4.1.2.1 Operations Based on Self-awareness

The enterprises significantly emphasize schedule planning while striving to balance production capacity and seasonal variations. Therefore, organizational development unfolds incrementally while adhering to a middle-path approach.

"We plan production schedules according to orders, considering the limited production capacity." (Enterprise A)

"Every day, there is a lot of organic waste such as food and manure waste generation from our firm activities. Therefore, we invested in biogas from organic waste to make electricity generation less dependent on fossil fuels. Excess manure, we applied for fertilizer and use in soil improvement." (Enterprise C)

"Last year, we installed metering to monitor department electricity consumption. We can utilize this data to show the hotspots and can apply for energy conservation." (Enterprise C)

4.1.2.2 Refraining from Excessive Debt

The study results indicate a conscientious approach by companies, especially small enterprises, toward financial risk management. The companies exercised caution when borrowing money, ensuring that the amount borrowed aligns with their ability to repay, thus mitigating the risk of overindebtedness.

"We expand our business based on our budget; we refrain from borrowing money from financial institutions. Our enterprise allocates 35% of total finance capital to further the organization's operations." (Enterprise A)

The case studies illustrate adopting operations in moderation based on self-awareness. Business development aligned with their organization's awareness indicates middle-path development, as demonstrated in the significance of the results in operations based on self-awareness practices. The strengths and weaknesses of the organizations were explored to understand their positions. The moderation factor aligns with the principles articulated by (Jitpetch, 2021), employing similar factors in managing businesses within community enterprises in Bangkok. This management approach prioritizes low-risk activities and emphasizes long-term profitability, reflecting a developmental concept similar to the one presented by (Pimki & Ridsomboon, 2014). Their findings support that organizations with moderation operations strengthen businesses in the long term.

4.1.3 Sharing

The sharing practice across members leads to human development in the community. Furthermore, sharing lessons learned with the public, including competitors, can benefit all stakeholders.

4.1.3.1 Training and Learning Center

Knowledge sharing was identified as a key success characteristic among the case studies. These enterprises trained stakeholders on their experiences and encouraged them to apply this knowledge to align with their lifestyles.

"We attach importance to opening as a learning center to encourage those interested to learn and develop further." (Enterprise B)

4.1.3.2 Networking and Trust-building

Collaboration and perseverance are fundamental factors in establishing network and trust among stakeholders.

"We are focused on creating good business partnerships and do not treat stakeholders as competitors. We share knowledge and work as a team. This approach can be used to support each other in times of crisis. The circle of partnership we developed is through trust." (Enterprise C)

"There is mutual assistance among partners covering both process development and sales. We are open to collaborating with research institutes to develop innovation together." (Enterprise B)

"We help each other within our organization whenever someone is in need. We believe we are on the same team and serve good quality products to customers. If other departments need help, we are willing to give them a hand." (Enterprise B)

"We do not discourage each other because we see the founder who always acts as a role model. She is humble enough to work with us even during difficult tasks." (Enterprise C)

The results indicate that stakeholders' trust is essential to building good partnerships. Trust is based on honesty and shared collaboration, which is evident in the case studies' shared significance regarding virtues. This finding aligns with the results from (Mongsawad, 2007) in that virtuous characteristics can increase stakeholder's trust and economic development. Moreover, this study's observation results highlight additional behaviors—especially the diligence and integration of knowledge obtained from local wisdom. As mentioned earlier, Table 6 presents the key success characteristics of agricultural enterprises.

4.2 Outputs from SEP Agricultural Enterprise

Based on the agricultural enterprise case studies, we obtained the findings of outputs from SEP community enterprises across two main dimensions. The outputs were all linked to significant input from the previous section and are consistent with SEP principles. The dimensions comprise food and energy sufficiency and increased stakeholder quality of life, as discussed below.

4.2.1 Resource Sufficiency

The agricultural enterprises prioritized the production of safe and locally sourced food. The primary emphasis in their empirical practices is on using local produce for internal activities, serving customers, and meeting consumption needs within the enterprise first. Subsequently, any surplus is processed and sold as fresh and

Table 6 Summary of key success characteristics in agricultural enterprises

	Enterprise A	Enterprise B	Enterprise C
Adaptation	New product development aligned with the stakeholder's context [1]. Examples: instant rice powder for elders in the community and using solar driers	Adopted technology transformation based on long-term foresight [1]. Examples: biochar production	Adopted technology transformation based on long-term foresight [1]. Examples: renewable energy investment
Moderation	Avoiding extreme investment, aligning the business fit with potential [1, 2] (Khaosod, 2021)	Avoiding extreme investment, aligning the business fit with potential [1]	Operating the organization based on self-awareness and continuous monitoring for improvement [1, 2] (Tipmongkol & Pooripakdee, 2022)
Sharing	Opening for knowledge sharing and motivating people to collaborate [1, 2]	Learning new knowledge by training, practicing, and sharing knowledge with stakeholders [1]	Building trust and good partnerships [1, 2] (Gray, 2022; Limpaiboon, 2018)

Note: Data retrieved from (1) case interviews and (2) public material/company websites

finished products, highlighting the strong commitment to food sufficiency within these firms.

> "We do not have to buy many foods from the market. We purchase only meat products and reduce family expenses for food. If we have excess eggs or vegetables, we can share them with our neighbors in the community." (Enterprise B)
>
> "We serve our food from our farms, which we grow by ourselves. I adopt planting from this location and apply it to cooking for my family using our own vegetables." (Enterprise C)

4.2.1.1 Energy Sufficiency

The case studies demonstrate energy sufficiency achieved by installing renewable energy sources and improving energy efficiency. Underpinned by a well-defined long-term plan, this approach seeks to harness self-generated energy. This mitigates risks linked to price volatility and the environmental impact of fossil fuel consumption.

> "We installed solar cells for water pumping, and we plan to expand solar cells to chicken farms to help reduce the dependency on conventional fuel." (Enterprise B)
>
> "Our energy comes from renewable as solar and biogas covering more than 90% of total energy consumption because we focus on the long-term operations. Moreover, we have food waste and cow and goat manure that can used as feedstock for biogas. So, we installed biogas to enhance the renewable energy capacity in our area." (Enterprise C)

The findings demonstrate that these enterprises focus on resource sufficiency combined with material management. They aim to source local materials and energy, reflecting their priority to strengthen self-reliance. This can lead to less dependency on imported energy and external materials while saving costs on imports and transportation, as shown in Enterprises B and C. This finding also supports the notion that the case study enterprises primarily rely on local resources.

4.2.2 Improving Stakeholder Quality of Life

The enterprises' concern regarding stakeholders, especially the impact on members and employees, was evident. Promoting work–life balance and job satisfaction was illustrated by the case studies, illustrating the key dimensions discussed in this section.

4.2.2.1 Work–Life Balance

The enterprises prioritized flexible working hours to facilitate a better balance between professional responsibilities and family care to enhance employee well-being. This emphasis is reflected in the enterprises' efforts toward improving the quality of life of the members' families.

> "Previously, I worked in the city. So I had to leave home early morning. I did not have enough time to prepare meals for my children. Since working for the enterprise, I have more time to care for my family. This has increased the happiness in our family." (Enterprise A)
>
> "The company allows the adjustment of working hours. This flexibility in caring for the family differs for each employee. For instance, when I just gave birth, working hours must be adjusted accordingly." (Enterprise C)

4.2.2.2 Self-satisfaction

The most significant characteristic expressed by the employees in this case study was life satisfaction. This was reflected in their subjective happiness and ability to support family.

> "Before working here, I worked as a worker in construction. Right now, I work on a plantation, and it makes me feel comfortable and happy to work here. I am responsible for taking care of my plantation area. The leader keeps encouraging and being humble to us, and I appreciate this." (Enterprise C)
>
> "I am enjoying it while I work here. I can do many challenges with the support of our teamwork. Here, we work with a good relationship, and we serve the good quality product for customers." (Enterprise C)

As mentioned earlier, the members and employees worried less about economic issues due to sufficient income. Household expenses decreased because they could plan their finances based on household accounting. They can consume food planted in their own house while generating revenue from excess products. Furthermore, they do not have to pay transportation fees to commute to urban centers for work. This demonstrates that income can be secure even during crises. The

Table 7 Summary of critical performance of agricultural enterprises

	Enterprise A	Enterprise B	Enterprise C
Resource sufficiency	Products developed and produced from community resources [1]	Prioritized using and circulating resource utilization within the organization itself [1]	Leveling up energy and material sufficiency and resource conservation [1, 2] (Gray, 2022)
Improving quality of life	Increasing family time and building members' self-esteem [1, 2] (Sansaiporpiang, 2013)	Increasing family time and providing food of good quality [1]	Increasing the life satisfaction of employees and stakeholder satisfaction [1]

Note: data retrieved from (1) case interviews and (2) public material

self-sufficiency regarding energy and food supply illustrates the impact at the family and community level that working for the SEP enterprises entails.

Moreover, employees who have more time to spend with family experience increased happiness and, thus, greater participation. This reduces family stress and enhances their pride in each other based on increased participation. This finding supports the research regarding how work duration relates to a family's happiness level (Wojewódzka-Wiewiórska et al., 2019). Therefore, if family engagement is more robust, it will improve quality of life. As mentioned earlier, Table 7 presents the critical performance of agricultural enterprises.

4.3 *Outcomes from SEP Agricultural Enterprises*

The case studies yielded significant outcomes, highlighting the positive impact of SEP on workplace well-being. This was particularly evident regarding stakeholder satisfaction and long-term performance, supported by resource sufficiency. Our results indicate that well-being, encompassing economic, social, and environmental dimensions, was observed in the workplace of the SEP agricultural enterprises examined in this case study. Factors such as work–life balance, quality of life, and job satisfaction contribute to subjective well-being. Moreover, our results reveal a link between resource sufficiency, characterized by limited material consumption, and the enhancement of well-being. This finding aligns with the principles of the "beyond GDP" and "degrowth" movements. For instance, recent studies (Fioramonti et al., 2022; Komatsu et al., 2022) emphasize that the sufficiency paradigm promotes a well-being economy. Thus, the observed outcomes of this research underscore the positive impact on well-being within agricultural enterprises.

5 Discussion

Based on a strengths and weaknesses analysis, we found that the limitation of workers in community enterprises was the main issue, consistent with the findings of (Naipinit et al., 2016). Therefore, enterprises need to shift the new generational attitude to convince upcoming workers of the tangible value of well-being, as indicated by the results. Additionally, moving from a short- to long-term focus is supported by perseverance and reasonableness, reflecting the main impacts regarding the positive characteristics of SEP.

5.1 Key Success Factors of SEP in Agricultural Enterprises

Three critical success factors are significant in the SEP integration within agricultural enterprises. First, the willingness to adapt is supported by past studies (Somwethee et al., 2023). Driven by organizational enrichment and skills, adaptability contributes to organizational sustainability. This adaptability is characterized by holistic thinking and the ability to tailor adaptation strategies to fit the organization's unique potential and context. The concerning impacts regarding self-immunity based on risk management and company responsibility motivate companies to adapt sustainably. This finding is consistent with previous research. For instance, (De Boon et al., 2024) demonstrated the factors related to the willingness to develop sustainable adaptation in the agricultural sector, which include risk appraisal and the outcome of activity. However, to encourage continued business cycle adaptation, monitoring and checking for improvements should be integrated into company processes.

Second, the application of the moderation mindset, often referred to as the "middle path," was evident in our study findings. Two highlighted behaviors we observed include conducting business operations based on self-awareness (Kansuntisukmongkol, 2017) and conducting business operations based on needs rather than wants. This finding supports the self-sufficiency of businesses, as stated in (Bocken & Short, 2016). Furthermore, our findings underscore the importance of avoiding overinvestment and excessive debt in mitigating financial risks amid rapid global change. A notable behavior observed across all case studies was implementing significant schedule planning to balance demand and supply, reflecting a moderation principle at the company level. Moreover, these enterprises focused on long-term benefits and invested according to their capacity to generate returns.

Third, this research finding suggests a noteworthy association between collaborative work relationships, sharing, and mutual assistance, which become apparent when employees perceive their work tasks as meaningful. This observation aligns with prior research indicating that enterprises fostering a sense of meaningful and cooperative work enhance workplace relationships (Freeman et al., 2021; Rothausen & Henderson, 2019). The evident behavior of sharing in this study, particularly knowledge sharing, affects workplace happiness. This finding aligns with those of (Sarapirom & Sarkar, 2018) in that sharing and altruism are factors that affect well-being and happiness. Additionally, sharing among enterprises can expand knowledge and build collaborative partnerships with stakeholders. These contribute to building relationships within society and trust among stakeholders. Past studies also support the notion of trust and reputation in a family business as essential for constructing the stakeholder relationship (Chaudhary et al., 2021). Previous research in the construction sector (Carvajal-Arango et al., 2021) also found that interpersonal relationships are significant drivers of workplace well-being. Thus, it is evident that the interconnection in organizations with the same values and sharing can foster a good workplace atmosphere. This indicates that knowledge combined with virtue as the foundation of SEP is essential.

Key SEP success factors for agricultural enterprises can be highlighted as adaptability, the moderation mindset, and collaborative work relationships while supported by sound knowledge and virtues, aligning the evidence from (Mettathamrong et al., 2023). Even though SEP companies in Thailand have no monetary incentives, the findings emphasize that companies apply SEP voluntarily, focusing on value based on long-term goals, which begins with the SEP mindset of the people in the company. These factors underscore the importance of SEP principles in business practices in contributing to sustainable and resilient enterprise development. While SEP enterprises may encounter friction or setbacks along the way, their adherence to these critical success factors enables them to withstand crises and emerge with resiliency. The COVID-19 pandemic serves as a testament to the resilience of SEP agricultural enterprises, as many could adapt their operations and survive amid unprecedented challenges.

5.2 Main Output of SEP in Agricultural Enterprises

This study identified two prominent outcomes of SEP within agricultural enterprises, each contributing to the overall well-being and sustainability of the enterprises examined in this case study.

First, resource sufficiency was evident in the findings, as dependency on internal resources was a priority for the enterprises in these case studies. Specifically, the enterprise case studies strived for self-reliance in energy consumption by increasing the proportion of renewable energy within their practices. Notably, shifting consumption from wants to needs can encourage organizations to reach higher levels of resource reliance. Moreover, the raw material enterprises intended to produce from their processes were prioritized based on self-awareness. This finding is supported by endogenous development theory by "development from within" based on the appropriation of a specific society. This notion illustrates the importance of well-being in communities (Nishikawa & Pimbert, 2022).

Second, the enterprise case studies contributed to upgrading stakeholder quality of life, reflecting the positive impact of SEP characteristics examined in this study.

The stakeholders whose quality of life has improved include employees, customers, and communities. This is evident when an enterprise aims to introduce products that promote good health, where value-based enterprises tend to increase quality along the supply chain. Similarly, the findings in this study indicate the balance between work and family, particularly within small-scale farming communities, as evident in enterprises with an SEP mindset. The findings confirm that SEP behaviors positively correlate with subjective well-being, surpassing the influence of household income (Poohongthong & Uwanno, 2009). This result is consistent with (Merkin, 2020) in that the life satisfaction of employees is reflected in their organizational citizenship behavior. Another reason for companies to strengthen employee well-being is the interconnections among people, which is clearly expressed in the culture of East Asia (Lee & Bae, 2022; Uchida & Rappleye, 2024).

Furthermore, if enterprises focus on stakeholder satisfaction, contentment among a vast array of stakeholders can support a sustainable vision (Petison & Kantabutra, 2022), which can then foster positive well-being. Stakeholder satisfaction includes the effect of organizational leaders acting as role models, encouraging virtuous behavior and a willingness to contribute to the organization, as supported by previous research (Elsamani et al., 2023; Ketprapakorn & Kantabutra, 2019a). Moreover, employees who are granted autonomy by their supervisors can achieve job satisfaction (Zhang et al., 2023). The results confirm earlier research stating that drivers of economic well-being include self-sufficiency, equity, and sustainability (Fioramonti et al., 2022). In the context of these case studies, managerial openness to employee opinions contributes to job satisfaction and a sense of value in the workplace. While our research focused on agricultural enterprises, these findings can apply to businesses across various industries.

The implications of SEP on agricultural enterprise's outputs are resource sufficiency and quality of life improvement. This suggests that addressing interconnections among economic, social, and environmental issues is essential for organizations practicing SEP, leading to workplace well-being. This aligns with the approach applied to business management in Bhutan, which emphasizes holistic management in leading to well-being and Gross National Happiness (Richardson, 2023).

Finally, this study constructed an agricultural enterprise well-being framework based on the case study results. The framework is based on the comprehensive thinking that is significant to the SEP principle, which leads to fostering workplace well-being. This emphasizes (O'Mahony, 2022) that sustainable well-being arises from the interconnectedness of value, environment, and societal dimensions. The framework, as depicted in Fig. 5, demonstrates sustainable development in a dynamic system that supports organizational well-being in the long run. The framework also reflects the flow from input to outputs and outcomes in agricultural enterprises. The critical process in this framework involves continual monitoring and modification to reach well-being in the workplace.

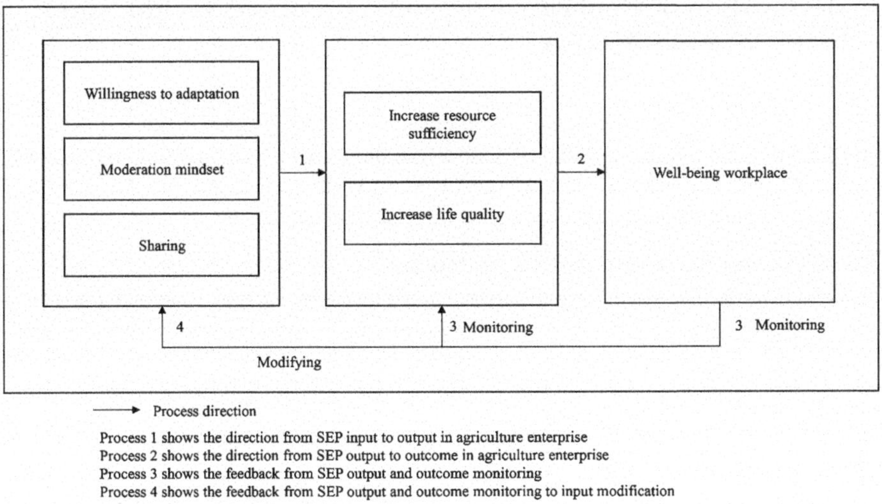

Fig. 5 Agricultural enterprise well-being framework

6 Conclusion and Limitations

This study illustrates the significant characteristics of adopting SEP in organizational business processes, encompassing both mindset and outcomes. The empirical results indicate that the moderation mindset, adaptation readiness, and sharing are key success characteristics of agricultural enterprises practicing SEP. These findings support stakeholder satisfaction and long-term performance in business.

Furthermore, the outcomes from enterprises' SEP activities support well-being across the workplace, society, and the environment. Companies that integrate SEP into their business processes can alter employee and stakeholders' attitudes within the value chain, fostering sustainability in business. However, monitoring and modifications are required to keep improving an organization to align it with applying SEP principles. The proposed framework in this study can be used to benefit other businesses, thereby enhancing the potential for sustainable development across various sectors.

The study's limitations include two issues that should be addressed in future research. First, this study was limited by a small number of case studies. Future research should expand the number of case studies to confirm the practicality of qualitative research. Moreover, various community enterprises should be included in the study sample to strengthen the robustness of the recommendations on business practices. Second, future research should conduct quantitative data analysis to measure subjective well-being to better understand the relationship between SEP behavior and workplace well-being.

7 Declaration of Generative AI and AI-assisted Technologies in the Writing Process

During the preparation of this work, the authors used Grammarly and ChatGPT to improve the grammar and readability and language of the manuscript. After using these tools and services, the authors reviewed and edited the content as needed and take full responsibility for the content of the publication.

CRediT Authorship Contribution Statement

Ruethai Onbhuddha: Conceptualization, methodology, formal analysis and writing—original draft. Seiichi Ogata: Conceptualization, supervision, writing—original draft.

Declaration of Competing Interest

The authors declare that they have no known competing financial interests or personal relationships that could have appeared to influence the work reported in this chapter.

References

Akkarangkul, P. (2019). Economic populism vs economic sufficiency: A crossroad facing Thailand. *NIDA Development Journal, 59*, 1–22.

Alkhodary, D. A. (2023). Exploring the relationship between organizational culture and well-being of educational institutions in Jordan. *Administrative Sciences, 13*, 92. https://doi.org/10.3390/admsci13030092

Aschemann-Witzel, J., Mulders, M. D., & Mouritzen, S. L. T. (2023). Outside-In and bottom-up: Using sustainability transitions to understand the development phases of mainstreaming plant-based in the food sector in a meat and dairy focused economy. *Technological Forecasting and Social Change, 197*, 122906. https://doi.org/10.1016/j.techfore.2023.122906

Bank of Thailand. (2024a). Labour force survey (new series ISIC Rev.4) (2011–2022). https://app.bot.or.th/BTWS_STAT/statistics/BOTWEBSTAT.aspx?reportID=638&language=eng.

Bank of Thailand. (2024b). Working conditions of the population classified by business type.

BCG's Research Community. (2018). White paper BCG in action.

Bekhet, A. K., & Zauszniewski, J. A. (2012). Methodological triangulation: An approach to understanding data. *Nurse Researcher, 20*, 40–43. https://doi.org/10.7748/nr2012.11.20.2.40.c9442

Bergsteiner, H., & Piboolsravut, P. D. (2016). The sufficiency economy philosophy process. In *Sufficiency thinking*. Allen & Unwin.

Bocken, N. M. P., & Short, S. W. (2016). Towards a sufficiency-driven business model: Experiences and opportunities. *Environmental Innovation & Societal Transitions, 18*, 41–61. https://doi.org/10.1016/j.eist.2015.07.010

Carè, R., Boitan, I. A., & Fatima, R. (2023). How do FinTech companies contribute to the achievement of SDGs? Insights from case studies. *Research in International Business & Finance, 66*, 102072. https://doi.org/10.1016/j.ribaf.2023.102072

Carvajal-Arango, D., Vásquez-Hernández, A., & Botero-Botero, L. F. (2021). Assessment of subjective workplace well-being of construction workers: A bottom-up approach. *Journal of Building Engineering, 36*, 102154. https://doi.org/10.1016/j.jobe.2021.102154

Chang, R.-D., Zuo, J., Zhao, Z.-Y., Zillante, G., Gan, X.-L., & Soebarto, V. (2017). Evolving theories of sustainability and firms: history, future directions and implications for renewable energy research. *Renewable & Sustainable Energy Reviews, 72*, 48–56. https://doi.org/10.1016/j.rser.2017.01.029

Chaudhary, S., Dhir, A., Ferraris, A., & Bertoldi, B. (2021). Trust and reputation in family businesses: A systematic literature review of past achievements and future promises. *Journal of Business Research, 137*, 143–161. https://doi.org/10.1016/j.jbusres.2021.07.052

de Boon, A., Sandström, C., & Rose, D. C. (2024). To adapt or not to adapt, that is the question. Examining farmers' perceived adaptive capacity and willingness to adapt to sustainability transitions. *Journal of Rural Studies, 105*, 103171. https://doi.org/10.1016/j.jrurstud.2023.103171

De Temmerman, J., Slabbinck, H., & Vermeir, I. (2023). The full package of package-free retail environments: A mixed methods study on multiple stakeholder perspectives of package-free shopping. *Sustainable Production & Consumption, 41*, 404–417. https://doi.org/10.1016/j.spc.2023.08.020

Elsamani, Y., Mejia, C., & Kajikawa, Y. (2023). Employee well-being and innovativeness: A multi-level conceptual framework based on citation network analysis and data mining techniques. *PLoS One, 18*, e0280005. https://doi.org/10.1371/journal.pone.0280005

FAO. 2023. *World food and agriculture – Statistical yearbook 2023*. Rome, Italy: Food and Agriculture Organization. https://doi.org/10.4060/cc8166en.

Fioramonti, L., Coscieme, L., Costanza, R., Kubiszewski, I., Trebeck, K., Wallis, S., Roberts, D., Mortensen, L. F., Pickett, K. E., Wilkinson, R., Ragnarsdottír, K. V., McGlade, J., Lovins, H., & De Vogli, R. (2022). Wellbeing economy: An effective paradigm to mainstream post-growth policies? *Ecological Economics, 192*, 107261. https://doi.org/10.1016/j.ecolecon.2021.107261

Freeman, R. E. (2010). *Strategic management: A stakeholder approach*. Cambridge University Press.

Freeman, R. E., Dmytriyev, S. D., & Phillips, R. A. (2021). Stakeholder theory and the resource-based view of the firm. *Journal of Management, 47*, 1757–1770. https://doi.org/10.1177/0149206321993576

Gomes, S., & Pinho, M. (2023). Can we count on the commitment of European SMEs to achieve SGD12? An exploratory study of business sustainability. *Journal of Cleaner Production, 425*, 139016. https://doi.org/10.1016/j.jclepro.2023.139016

Gray, G. (2022). Agritourism with a mission. In *Thailand's BCG transformation: 40 case studies on the bio-circular-green strategy and the sufficiency economy philosophy in action* (pp. 118–121). NIDA's Sustainable Development and Sufficiency Economy Studies Center.

Harsanto, B., Mulyana, A., Faisal, Y. A., & Shandy, V. M. (2022). Open innovation for sustainability in the social enterprises: An empirical evidence. *Journal of Open Innovation, 8*, 160. https://doi.org/10.3390/joitmc8030160

Hayles, C. S., & Dean, M. (2015). Social housing tenants, climate change and sustainable living: A study of awareness, behaviours and willingness to adapt. *Sustainable Cities & Society, 17*, 35–45. https://doi.org/10.1016/j.scs.2015.03.007

Isham, A., Mair, S., & Jackson, T. (2021). Worker wellbeing and productivity in advanced economies: Re-examining the link. *Ecological Economics, 184*, 106989. https://doi.org/10.1016/j.ecolecon.2021.106989

Jiang, Y., Ni, H., Guo, X., & Ni, Y. (2023). Integrating ESG practices and natural resources management for sustainable economic development in SMEs under the double-carbon target of China. *Resources Policy, 87*, 104348. https://doi.org/10.1016/j.resourpol.2023.104348

Jitpetch, P. (2021). Management of community enterprises by applying the sufficiency economic philosophy effectively. *Journal of Management Sciences, 8*, 39–49. https://doi.org/10.14456/jmsnpru.2021.4

Jitsuchon, S. (2019). Thailand's sufficiency economy philosophy as an alternative path to sustainable development. *European Journal of Sustainable Development, 8*, 191. https://doi.org/10.14207/ejsd.2019.v8n2p191

Kansuntisukmongkol, K. (2017). Philosophy of sufficiency economy for community-based adaptation to climate change: Lessons learned from Thai case studies. *Kasetsart Journal of Social Sciences, 38*, 56–61. https://doi.org/10.1016/j.kjss.2016.03.002

Kantabutra, S. (2019). Achieving corporate sustainability: Toward a practical theory. *Sustainability, 11*, 4155. https://doi.org/10.3390/su11154155

Kantabutra, S., & Punnakitikashem, P. (2020). Exploring the process toward corporate sustainability at a Thai SME. *Sustainability, 12*, 9204. https://doi.org/10.3390/su12219204

Kantabutra, S., & Siebenhuner, T. (2011). Predicting corporate sustainability: A Thai approach. *Journal of Applied Business Research, 27*, 123. https://doi.org/10.19030/jabr.v27i6.6471

Ketprapakorn, N., & Kantabutra, S. (2019a). Culture development for sustainable SMEs: Toward a behavioral theory. *Sustainability, 11*, 2629. https://doi.org/10.3390/su11092629

Ketprapakorn, N., & Kantabutra, S. (2019b). Sustainable social enterprise model: Relationships and consequences. *Sustainability, 11*, 3772. https://doi.org/10.3390/su11143772

Khaosod. (2021). Community enterprise Ban Ku Daeng farmers promoting processed herbal products. WWW document. https://www.khaosod.co.th/newspaper/news_5653951

Komatsu, H., Rappleye, J., & Uchida, Y. (2022). Is happiness possible in a Degrowth society? *Futures, 144*, 103056. https://doi.org/10.1016/j.futures.2022.103056

Korphaibool, V., Chatjuthamard, P., & Treepongkaruna, S. (2021). Scoring sufficiency economy philosophy through GRI standards and firm risk: A case study of Thai listed companies. *Sustainability, 13*, 2321. https://doi.org/10.3390/su13042321

Kwangsawad, A., & Jattamart, A. (2022). Overcoming customer innovation resistance to the sustainable adoption of chatbot services: A community-enterprise perspective in Thailand. *Journal of Innovation & Knowledge, 7*, 100211. https://doi.org/10.1016/j.jik.2022.100211

Lee, J.-E., & Bae, S.-M. (2022). The relationship between social capital, the happiness of others, and the happiness of Korean adults: A national sample study. *Psychiatry Investigation, 19*, 958–964. https://doi.org/10.30773/pi.2022.0050

Lennartz, J., Toxopeus, M. E., & Van Der Meulen, J. (2022). Analysis of environmental transitions for tool development. *Procedia CIRP, 105*, 799–804. https://doi.org/10.1016/j.procir.2022.02.132

Limpaiboon, A. (2018). A large hotel rich in sustainability. WWW document. *The Cloud*. https://readthecloud.co/porlaewdee-sivatel/. Accessed 3 Dec 2024.

Merkin, R. S. (2020). Employee life satisfaction and social-capital factors relating to organizational citizenship. *Performance Improvement Quarterly, 33*, 55–75. https://doi.org/10.1002/piq.21317

Methamontri, Y., Tsusaka, T. W., Zulfiqar, F., Yukongdi, V., & Datta, A. (2022). Factors influencing participation in collective marketing through organic rice farmer groups in northeast Thailand. *Heliyon, 8*, e11421. https://doi.org/10.1016/j.heliyon.2022.e11421

Mettathamrong, J., Upping, P., & Deeudom, S. (2023). Approach to applying sufficiency economy philosophy in community enterprise management towards sustainability. *Sustainability, 15*, 5338. https://doi.org/10.3390/su15065338

Ministry of Foreign Affairs, Kingdom of Thailand. (2017). *Sufficiency economy philosophy: Thailand's path towards the sustainable development goals*. MFA.

Mongsawad, P. (2007). Sufficiency economy philosophy and economic management. *NIDA Development Journal, 47*, 159–179.

Mongsawad, P., & Thongpakde, N. (2016). Sufficiency economy philosophy: A holistic approach to economic development and mainstream economic thought. *Asian Social Science, 12*, 136. https://doi.org/10.5539/ass.v12n7p136

Naipinit, A., Sakolnakorn, T. P. N., & Kroeksakul, P. (2016). Strategic management of community enterprises in the upper northeast region of Thailand. *Journal of Enterprising Communities: People & Places in the Global Economy, 10*, 346–362. https://doi.org/10.1108/JEC-06-2015-0032

NESDC, The Office of the National Economic and Social Development Council. (2022). The thirteen national economic and social development plan (2023–2027).

Nguyen, H., Pham, A. V., Pham, M. D., & Pham, M. H. (2023). Business resilience: Lessons from government responses to the global COVID-19 crisis. *International Business Review, 32*, 102166. https://doi.org/10.1016/j.ibusrev.2023.102166

Nishikawa, Y., & Pimbert, M. (Eds.). (2022). *Seeds for diversity and inclusion: Agroecology and endogenous development*. Springer International Publishing. https://doi.org/10.1007/978-3-030-89405-4

Norris, S. (2024). In the eye of the beholder: Stakeholder perceived value in sustainable business models. *Long Range Planning, 57*, 102406. https://doi.org/10.1016/j.lrp.2023.102406

O'Mahony, T. (2022). Toward sustainable wellbeing: Advances in contemporary concepts. *Frontiers in Sustainability, 3*, 807984. https://doi.org/10.3389/frsus.2022.807984

OECD. (2022). *Financing SMEs and entrepreneurs 2022: An OECD scoreboard, full country profile: Thailand*. OECD Publishing. https://doi.org/10.1787/e9073a0f-en.

OECD. (2023). *Valuing business impacts in the areas of wage inequality and employee well-being (OECD Papers on Well-being and Inequalities No. 15), OECD Papers on Well-being and Inequalities*. OECD Publishing. https://doi.org/10.1787/740deb2f-en.

OSMEP (2022) Office of small and medium enterprises promotion. Annual report 2021. WWW document. https://en.sme.go.th/en/. Accessed 3 Dec 2024.

Oyelami, O., & Olivier, M. (2015). Using Yin's approach to case studies as a paradigm for conducting examinations. In G. Peterson & S. Shenoi (Eds.), *IFIP advances in information & communication technology, advances in digital forensics* (Vol. XI, pp. 45–59). Springer International Publishing. https://doi.org/10.1007/978-3-319-24123-4_3

Petison, P., & Kantabutra, S. (2022). A quest for a sustainable social enterprise model: The case of Amphawa Chaipattananurak, the Kingdom of Thailand. *Sustainability, 15*, 326. https://doi.org/10.3390/su15010326

Pimki, T., & Ridsomboon, J. (2014). The practical use of the royal sufficiency economy with community enterprise in Chanthaburi Province. *SDU Research Journal, 10*, 1–21.

Poohongthong, C., & Uwanno, T. (2009). Relationship between living with sufficiency economy philosophy and subjective well-being in farmer. *NIDA Development Journal, 49*, 147–165.

Rasiah, R., & Ren, Y. (2023). Sustainable management of a leading Chinese telecommunication multinational: A case study of company X in host country Malaysia. *Cleaner and Responsible Consumption, 8*, 100092. https://doi.org/10.1016/j.clrc.2022.100092

Richardson, J. A. (2023). How does gross national happiness offer an integrated perspective linked with health, economics, and nature? *Journal of Ayurveda & Integrative Medicine, 14*, 100601. https://doi.org/10.1016/j.jaim.2022.100601

Rothausen, T. J., & Henderson, K. E. (2019). Meaning-based job-related well-being: Exploring a meaningful work conceptualization of job satisfaction. *Journal of Business & Psychology, 34*, 357–376. https://doi.org/10.1007/s10869-018-9545-x

Ruiz-Pérez, F., Lleo, A., & Ormazabal, M. (2021). Employee Sustainable behaviors and their relationship with corporate sustainability: A Delphi study. *Journal of Cleaner Production, 329*, 129742. https://doi.org/10.1016/j.jclepro.2021.129742

Sansaiporpiang. (2013). Learning the principles of work and applying them to groups of farmers to be strong and sustainable. WWW document. https://sansaiporpiang.mju.ac.th/wtms_newsDetail.aspx?nID=10404&lang=th-TH. Accessed 3 Dec 2024.

Sarapirom, K., & Sarkar, H. (2018). Study of sufficiency economy philosophy and its impact on individuals, communities and organizations. *Interdisciplinary Research Review, 13*, 8–12. https://doi.org/10.14456/jtir.2018.2

Sepasgozar, S. M. E., & Davis, S. (2018). Construction technology adoption cube: An investigation on process, factors, barriers, drivers and decision makers using NVivo and AHP analysis. *Buildings, 8*, 74. https://doi.org/10.3390/buildings8060074

Silva, M. E., Silvestre, B. S., Del Vecchio, R. C., & Ponte, and José E. O. Cabral. (2021). Managing micro and small enterprise supply chains: A multi-level approach to sustainability, resilience and regional development. *Journal of Cleaner Production, 311*, 127567. https://doi.org/10.1016/j.jclepro.2021.127567

Somwethee, P., Aujirapongpan, S., & Ru-Zhue, J. (2023). The influence of entrepreneurial capability and innovation capability on sustainable organization performance: Evidence of community enterprise in Thailand. *Journal of Open Innovation, 9*, 100082. https://doi.org/10.1016/j.joitmc.2023.100082

Song, H.-C. (2020). Sufficiency economy philosophy: Buddhism-based sustainability framework in Thailand. *Business Strategy & the Environment, 29*, 2995–3005. https://doi.org/10.1002/bse.2553

Songsom, A. (2014). Structural equation model of small and medium enterprise management based on the philosophy of sufficiency economy. *WMS Journal of Management, 3*, 51–56.

Supasri, T., Gheewala, S. H., Macatangay, R., Chakpor, A., & Sedpho, S. (2023). Association between ambient air particulate matter and human health impacts in northern Thailand. *Scientific Reports, 13*, 12753. https://doi.org/10.1038/s41598-023-39930-9

The Thailand Enterprise Amounts from 201–2022 [WWW document]. (2022). Off. SMEs Promot. https://www.sme.go.th/. Accessed 2 June 2022.

Tipmongkol, P., & Pooripakdee, S. (2022). Narrative of success on agro-tourism management with sufficiency economy philosophy for sustainability at Rai Ruen Rom organic farm, Thoeng District, Chiangrai Province. *Valaya Alongkorn Review: Research on Humanities & Social Sciences, 12*, 135–150.

TISI, Thai Industrial Standard. (2013). Guidance on sufficiency economy for industries.

Uchida, E., & Kino, Y. (2021). Study on the relationship between employee satisfaction and corporate performance in Japan via text mining. *Procedia Computer Science, 192*, 1730–1739. https://doi.org/10.1016/j.procs.2021.08.178

Uchida, Y., & Rappleye, J. (2024). *An interdependent approach to happiness and well-being*. Springer International Publishing. https://doi.org/10.1007/978-3-031-26260-9

Wojewódzka-Wiewiórska, A., Kłoczko-Gajewska, A., & Sulewski, P. (2019). Between the social and economic dimensions of sustainability in rural areas – In search of farmers' quality of life. *Sustainability, 12*, 148. https://doi.org/10.3390/su12010148

Zhang, J., Huang, R., Chen, Q., & Zhao, G. (2023). The relationships between supervisor-subordinate Guanxi, perceived supervisor autonomy support, autonomous motivation, and employee job satisfaction: Evidence from international hotel chains in China. *International Journal of Hospitality Management, 108*, 103354. https://doi.org/10.1016/j.ijhm.2022.103354

The Current Dilemma of the Social Healthcare System in Chile: An Analysis of the Current Situation and Discussion of the Pending Reforms, from a Social Sustainability Management Viewpoint

Fernando Yanine

Contents

F. Yanine (✉)
Faculty of Engineering, Universidad Finis Terrae, Santiago, Chile
e-mail: fyanine@uft.cl

S. S. Muthu (ed.), *Assessment of Social Sustainability Management in Various Sectors*, Environmental Footprints and Eco-design of Products and Processes,
https://doi.org/10.1007/978-3-031-81029-9_5

1 Introduction

Social sustainability refers to the proactive identification and management of business impacts on society (Ajmal et al., 2018; Foot & Ross, 2017). Whether we talk about employees, workers in an industry's value chain, customers, and/or local communities, the goal should be to manage the means and resources—especially the environment—to ensure that current and future generations can thrive in a healthy, fair, and equal environment. Among good examples of social sustainability, we can cite healthcare, where efforts are aimed at ensuring that healthcare is accessible to all community members, regardless of their financial capabilities. Ensuring such a basic yet monumentally important right is absolutely fundamental to promoting social sustainability. There is no doubt that such a public goods goal would benefit everyone in society, not just those who pay for it (Talbot & Verrinder, 2017), as is the case of healthcare disparities in Chile. Some good examples of this effort to bring about effective and efficient public policy practices were the reforms made to the Chilean healthcare system back in the 1980s (Rotarou & Sakellariou, 2017; Barrientos, 2000) and, most importantly, the pending reforms that are still being discussed as of today, as the population faces the dire situation of Chile's healthcare system (Charney et al., 2021), especially with the precarious state of the National Health Fund (FONASA) and several individual and family private healthcare insurance companies like Social Security Health Institutions (ISAPREs). The main past reforms included the following: (1) the creation of FONASA, (2) the creation of the National Health Services System (SNSS), (3) the creation of ISAPREs, and (4) the decentralization and municipalization of primary healthcare (Martinez-Gutierrez & Cuadrado, 2017; Homedes & Ugalde, 2005). However, there are still many issues pending as there are those who want to move toward a more sustainable, personalized, and decentralized healthcare system, as some studies suggest (Núñez et al., 2020; Martinez-Gutierrez & Cuadrado, 2017), while the government stubbornly insists on a universal, one-fits-all national healthcare system despite the perilous state of the healthcare units and hospitals that are clearly insufficient to provide adequate service to the long waiting lists that keep growing in some parts of Chile (Núñez et al., 2020).

In this chapter, we discuss this situation and analyze the various factors and parties involved, and we also undertake a critical analysis of the current push for reforms from a social sustainability management viewpoint and what the outlook of the healthcare model is in regard to the social sustainability of the Chilean healthcare system overall. We also examine the main trends and challenges that the sector faces today and in the years to come.

1.1 A Brief Diagnosis of the Current Healthcare System in Chile and Its Implications

In regard to healthcare settings, and especially when referring to public healthcare systems, social sustainability relates to the ability of healthcare facilities, such as hospitals, clinics, and other types of healthcare systems, to enhance the quality of life and improve well-being in the population (Núñez et al., 2020). In light of this, we may safely ascertain, without hesitation, that healthcare in Chile is in a critical situation, as evidenced by the long waiting lists in public healthcare providers where hundreds of people wait for inter-consultation with a specialist or for access to a planned surgery, and the crisis in the system of private health institutions (ISAPREs) continues to deteriorate (Letelier, 2014; Crespo, 2018; Rørdam, 2016).

Likewise, this chapter sets out to analyze the difficult situation of private health insurers— the ISAPREs—and how it has evolved overtime, exploring its causes and the consequences it has entailed. As a starting point, for a comprehensive understanding of the problem, it is important to first make a diagnosis of the current state of the healthcare services' sector. Chile's healthcare system has been a subject of extensive debate, particularly in terms of its ability to deliver quality service and achieve sustainable performance (de la Jara & Bossert, 1995; Dominguez-Cancino et al., 2020; Bossert & Villalobos Dintrans, 2020). The dichotomy between the public and private healthcare sectors highlights significant disparities, necessitating a critical examination, from a social sustainability management viewpoint, that requires going back to the beginning of the healthcare sector reforms, in order to comprehend the full breadth of the crisis. This chapter aims to dissect these issues and propose viable reforms to enhance both the quality and sustainability of healthcare services in Chile.

1.2 The Dichotomy of Chile's Private and Public Healthcare Systems: Balancing the Challenges of Public Healthcare with Freedom of Choice

Indeed, Chile's healthcare system is at a critical juncture. As a country that has made significant strides in economic development, the disparity between its public and private healthcare services has become a prominent issue. The dichotomy of the two coexisting systems is ever more present and poignant, wherein government and the legislature must balance the challenges of providing affordable public healthcare with the freedom of choice for those in the private healthcare system. The hardships of those who are in the public healthcare system have deepened, and resentment and frustration have grown larger over the years while the private healthcare system, composed of the health insurance companies (ISAPREs), has become less accessible in terms of healthcare coverage and cost of the health insurance plans. These and the rising unemployment, due to very sluggish economic growth in 2023 and in the course of 2024, have led thousands of Chileans to seek healthcare service in the public system FONASA. However, this option—forced by the circumstances—has the downside of endless waiting periods for a specialty physician's appointment and lack of adequate medical infrastructure, along with other issues that have played against the possibility of having a timely, quality healthcare service.

All these drawbacks and hardships that are quite familiar to all those who are in the public healthcare system, along with several other issues, have led to unequal access to quality healthcare services, creating a rift between the two systems in terms of access to quality and timely healthcare and opportunity cost. Over the years, and especially during the last 4 years with the COVID-19 pandemic in 2020–2021 and the healthcare crisis in the public healthcare system FONASA, there is a pressing need for reforms that can bridge this gap and ensure sustainable, equitable healthcare for all citizens (Bossert & Villalobos Dintrans, 2020). Upon examining the current challenges faced by Chile's healthcare system and exploring the pending reforms through the lens of social sustainability management, we realize that while much money has been appropriated for public health and prevention programs, over the past three decades, the management of public healthcare programs and facilities, such as hospitals and municipal healthcare units, has been deficient and, in some instances, negligent (Alvial et al., 2021). Hence, the focus of this current analysis will be on how the healthcare system can deliver quality service while maintaining sustainable performance, ensuring that every Chilean has access to the healthcare they need.

1.3 Social Sustainability and Public Healthcare Systems: Connecting the Dots in the Chilean Context

Social sustainability is a critical concept that intertwines with the healthcare system, emphasizing the need to create and maintain healthy, equitable, and resilient communities (Cristiano et al., 2021). In the context of Chile's public healthcare system, social sustainability focuses on ensuring that the healthcare infrastructure not only meets the current needs of the population but also remains robust and accessible for future generations. However, the reality of the public healthcare system says otherwise. Hence, for social sustainability to effectively connect to a sustainable public healthcare system in Chile that can provide quality, affordability, and timely services, the following principles must be observed:.

1.3.1 Social Sustainability Principle of Equity and Access to Healthcare

When examining the social sustainability principle in the healthcare context, we realize that the current state of affairs in terms of the availability of quality and affordable healthcare for the population in Chile is dire. While this principle calls for ensuring equitable access to resources and opportunities for all members of society, this is farthest from the truth in the current healthcare system of Chile. A socially sustainable healthcare system must provide equitable access to medical services regardless of socioeconomic status, something that clearly is lacking in the current situation that Chileans face when seeking healthcare in the public healthcare system FONASA (Núñez et al., 2020). In Chile, this means addressing the disparity between the public and private sectors, ensuring that all citizens, particularly those relying on public healthcare, have access to high-quality medical care. While policy efforts have been made over the years to address such disparity, the results have been, for the most part, ineffective and, sometimes, outright deplorable. There is no doubt that a complete overhaul of the current public healthcare system in Chile is warranted, yet the powers in charge have mighty political forces against this badly needed system reengineering effort, especially when you look at the very large syndicates politically active as a mounting influence on public health policies, as well as the not less powerful college of physicians, the college of nurses, and other healthcare technical personnel.

While historical healthcare reforms in Chile that took place in the 1980s focused mainly on expanding healthcare coverage, improving infrastructure in underserved areas, and reducing waiting times in public hospitals, all of which are essential; the changes implemented unfortunately generated several disparities between the private and public healthcare systems that grew overtime. The measures implemented through the reforms helped, in part, bridge the gap between different population groups, fostering a more inclusive and fair healthcare system, but poor healthcare management in public healthcare facilities like hospitals and other healthcare units and lack of government oversight deepened the crisis and exacerbated the inequities between the two systems. Therefore, over time, the disparities continued to widen, and the inequities between those who could afford private healthcare in the ISAPREs and those who could not grew larger and larger, finally arriving to the dire situation that we see today. Special mention among the causes that were responsible for the demise of the public healthcare system were the lack of good professional management in the country's overall public health system and the lack of government oversight, leaving too much leeway to the system administrators and the powerful syndicates operating within the public health sector (Göttems & Mollo, 2020, 19; Gavurova et al., 2021). Thus, public healthcare facilities became ever more inefficient and insufficient in terms of their management and their capacity to provide service to a growing demand from the much larger sector of the population that had as their only option, the public healthcare system FONASA.

1.3.2 Some Historical Context: How the Decentralization of the Chilean Healthcare System Came About

According to data from the Department of Financial Studies of Chile's Ministry of Finance (1959), contained in the "*Manual de Organización del Gobierno de Chile. Santiago de Chile*," Chile's National Medical Service for Employees (SERMENA), which operated from 1942 to 1979, provided medical and dental care for public and private employees under an administrative or free choice system (where people could opt to buy "bonos" to get free choice healthcare) system, unlike the National

Health Service (SNS) where people had no choice but to accept whichever physicians and appointment dates were available, when they were available. However, very often people had to wait for a specialist and for a slot in their busy schedule, just as it is the case today, with the current national healthcare system FONASA.

SERMENA was created in 1942 by the merger of other healthcare systems in existence at the time, the health departments of the Caja Nacional de Empleados Públicos y Periodistas (CEPP) and the Caja de Previsión de Empleados Particulares (EMPART). The system provided healthcare for workers and for the rest of the country's population. Later came the Law 2763, which was adopted at the end of 1979, which redefined the structure of the public sector. It fused the then existing national healthcare system (Servicio Nacional de Salud) and SERMENA into a brand new single body, that is, the National Health Services System (SNS) in 1979, where the National Health Fund (FONASA)—which is in existence today—assumed the modality of free medical choice (Manuel, 2002).

As Annick Manuel (2002) correctly points out in his paper "*The Chilean Health System: 20 Years of Reforms*," the foundations of the current healthcare system of Chile were designed and implemented during the Pinochet era. The democratic regimes that followed introduced some changes but chose to preserve the core of the system (Manuel, 2002). Although many changes were positive in the sense that helped to improve the quality and efficiency of public healthcare services with the achieved decentralization, there were also serious downsides which we have already discussed, that have deepened over the last three decades. The healthcare reforms during the Pinochet government period and the post-Pinochet era, which are shown in Figs. 1 and 2, allowed the ISAPREs to emerge as the private healthcare option of

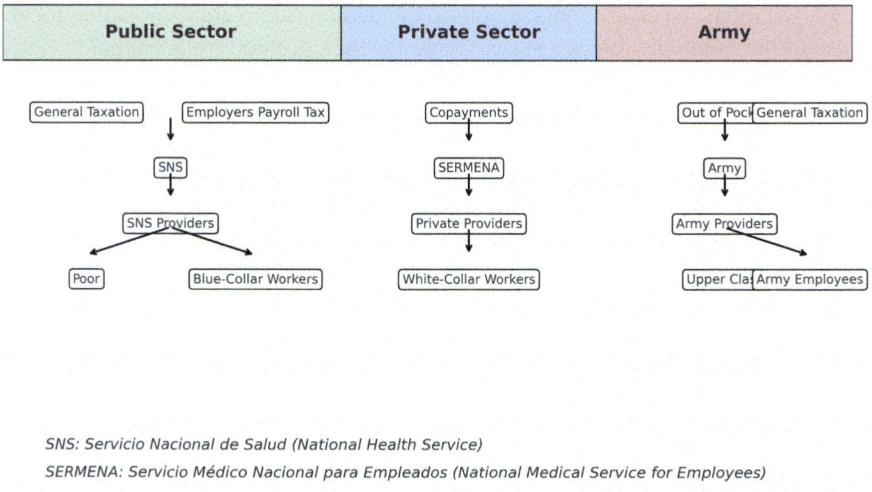

Fig. 1 A diagram depicting how the old national healthcare system of Chile worked. (Source: Manuel, 2002)

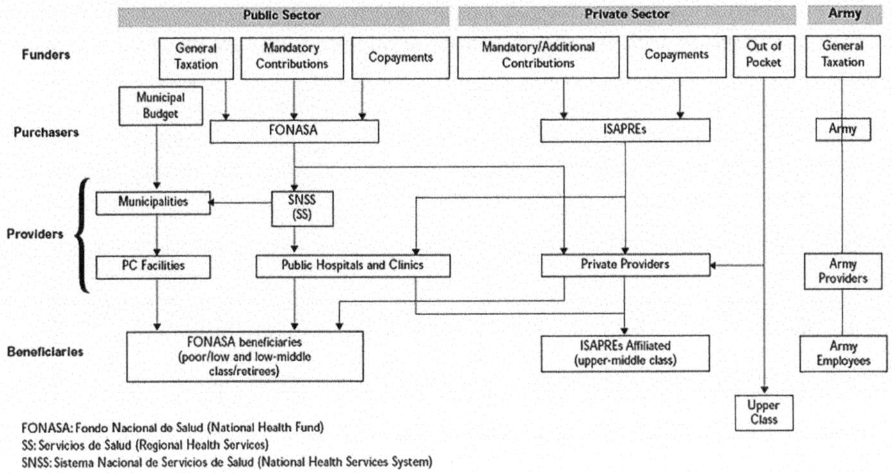

FIGURA 2. CHILEAN HEALTH CARE SYSTEM IN THE PINOCHET AND POST-PINOCHET ERA

Fig. 2 Diagram depicting Chile's healthcare system in the Pinochet government period and the post-Pinochet era. (Source: Manuel, 2002)

choice maintaining the coexistence of the public and private systems, while reinforcing the gap between the two systems, which we have already discussed.

As shown in Figs. 1 and 2, we can see the healthcare transformation fueled by the reforms carried out during the Pinochet regime. These reforms aimed basically at the decentralization of the public healthcare system and the development of the private healthcare industry (Manuel, 2002).

1.3.3 Social Sustainability Principle of Quality and Affordability of Healthcare Services

Maintaining high standards of living and well-being for all community members is essential in any modern society; no doubt, Chile aspires to achieve this fundamental principle in terms of healthcare. Quality healthcare services are fundamental to the well-being of individuals and communities, and we should expect that a sustainable healthcare system would be able to ensure that services are not only available and affordable, but also of high quality, leading to better health outcomes. However, this is not the case in Chile, and there are good reasons for this. Implementing standardized protocols, investing in continuous professional development for healthcare workers, putting the right people in charge of the public healthcare administration, and adequately governing the entire system from a political and managerial viewpoint are all essential tenets. Especially so when there are powerful political syndicates and non-government organizations (NGOs) which are strongly against the much-needed modernization of a decaying public healthcare system and a private healthcare alternative whose outlook seems gloomy and utterly uncertain. While it is true that there are well-managed public healthcare facilities in Chile and well-staffed and highly modern in terms of unfractured, having adopted advanced medical technologies that can enhance the quality of care in public hospitals and clinics, these are far from the norm. Ensuring consistent quality across the country is crucial for building trust in the public healthcare system, yet upon diagnosing the overall state of the Chilean public healthcare system, we come to realize that those high standards are not met (Núñez et al., 2020).

1.3.4 Social Sustainability Principle of Healthcare Affordability

Promoting economic stability and reducing financial barriers to essential services are essential when seeking to offer quality healthcare services that are also affordable for all segments of the population. High costs can lead to inequities in access and deter people from seeking necessary care, negatively impacting public health, which is unfortunately the case in Chile today. Subsidizing healthcare costs for low-income families, implementing sliding scale payment systems based on income, and negotiating better pricing for medications and treatments can make healthcare more affordable. This financial support is vital for reducing out-of-pocket expenses and ensuring that healthcare is accessible to everyone.

1.3.5 Social Sustainability Principle of Timely Healthcare Services

Ensuring that community needs are met promptly and effectively is also essential anywhere, and good government must strive to achieve this. Hence, in the context of public health policy, timely access to healthcare services is essential for preventing and managing health conditions and pandemics. Delays in receiving care can lead to worsening health outcomes and increased long-term costs and even deaths. Well, unfortunately that is the case in Chile with the public healthcare system FONASA, where several thousands of people desperately wait every year and many hundreds die waiting for healthcare (Peña & Silva, 2022; Bitran, 2013). Streamlining administrative processes, increasing the number of healthcare providers, and leveraging telemedicine to reach remote areas can help reduce wait times and improve the timeliness of care, but it is not enough. A major overhaul of the overall public healthcare system management must also be in place to ensure timely and effective, good quality healthcare. There is no way out of this mess unless major and drastic changes are made; otherwise, there will never be a sustainable healthcare system, where truly efficient service delivery systems critical for addressing urgent health needs and improving overall health outcomes become the norm.

2 Current Situation of the Healthcare System in Chile: How Did We Get Here?

According to the World Health Organization, health policy frameworks (2023) provide the context for implementation of programs aimed at aligning a country's priorities with its population's health needs, in partnership with government, health and development partners, civil society, and the private sector for improved use of available resources. Several reforms were implemented at the beginning of the 1980s by the military government of Augusto Pinochet, which would change the old and inefficient national healthcare system that existed in Chile until then and provide a new structure and functioning of the healthcare sector in the country. Among the main changes were the creation of the National Health Fund; the National Health Services System; the ISAPREs, which were the private healthcare insurance institutions to which people had the chance to affiliate themselves; and FONASA, where people that could not afford the ISAPREs could affiliate to this, the public healthcare system that, in theory, provided the same coverage but that, in reality, was much more vulnerable and deprived in terms of its possibilities, especially in regard to surgeries and consultation with specialists. And while the process of municipalization of primary healthcare was a good thing, it proved to be insufficient in terms of coverage and availability of infrastructure and resources, especially in the larger and more underprivilege communes of the country.

Among the virtues of the reforms implemented in the beginning of the 1980s are, on the one hand, the separation of functions between financing and healthcare and the decentralization processes that were put in place. Nevertheless, the current structure of the healthcare system has proven insufficient and unjust when one looks at the services and options that ISAPREs' members have as compared to what FONASA's affiliates have. Thus, the healthcare system that was created in Chile because of these reforms has produced and continues to produce serious inequities and inefficiencies in the overall management of the healthcare sector (Bossert & Villalobos Dintrans, 2020). One of the main characteristics of the current system's pitfalls lies in the fact that health contributions can be paid by the healthcare system, yet there are two health insurance systems that operate in parallel and with very different rationales. That is, contributors can opt for the public health insurance (FONASA) or the private health insurance offered by ISAPREs. Thus, Chile's healthcare system is divided into two main sectors: public and private. The public sector is managed by the National Health Fund (FONASA), which provides coverage to approximately 80% of the population, including the most vulnerable and lower-income groups. The private sector, on the other hand, is managed by the private health insurance institutions, termed ISAPREs, and serves around 20% of the population, primarily those with higher incomes.

FONASA operates through a network of public hospitals and clinics, offering services that range from primary care to specialized treatments. Despite the comprehensive coverage, the public healthcare system is often criticized for its long waiting times, overcrowded facilities, and variable quality of care. In contrast, the private sector boasts shorter waiting times and higher-quality services, but at a higher cost, making it less accessible to the average Chilean.

According to Bossert and Villalobos Dintrans (2020), many proposals over several administrations have been offered to address the key problems of the health system that was created by the Pinochet regime: inefficiency, segmentation, inequality, and lack of solidarity (Villalobos Dintrans, 2020). This two-tiered system involves a large publicly financed social insurance fund (FONASA) that covers roughly 80% of the population. FONASA funds a large network of public hospitals and primary care clinics, covering private healthcare under its "free choice" option (Bossert & Villalobos Dintrans, 2020). According to Zurynski et al., (2022) increasing healthcare costs and increasing demand for services and patient multimorbidity challenge the sustainability of healthcare systems. In the case of Chile, these challenges are all present today and have persisted over several decades, deepening the crisis and generating a rift between public and private healthcare systems, as evidenced by the global pandemic of 2020–2021. An overview of the literature teaches us that there are several built-in barriers to sustainable healthcare; often times, healthcare system's facilitators of healthcare programs do not always ensure system sustainability (Zurynski et al., 2023) Again, citing the same authors, in their extensive study entitled "How can the healthcare system deliver sustainable performance? A scoping review," they concluded that in order to implement long-lasting and effective change and to respond to new challenges, context-relevant definitions and frameworks are necessary, as well as having robust, flexible, and feasible measures to support the long-term sustainability and performance of healthcare systems (Zurynski et al., 2022).

2.1 Key Challenges and Issues Ahead: How Do We Get Out of This Mess?

The primary challenges facing Chile's healthcare system include the following:

1. *Inequality in Healthcare Access*: The division between public and private healthcare creates a stark disparity in access to services. Those reliant on the public system often face long waits and limited resources, while those who can afford private insurance enjoy prompt and superior care.
2. *Quality of Service*: The quality of healthcare services in the public sector is inconsistent, with significant variations in the standard of care provided across different regions and facilities. This inconsistency can lead to adverse health outcomes for patients who depend solely on public healthcare.
3. *Financial Sustainability*: Funding and resource allocation in the public healthcare system are critical issues. Public hospitals and clinics frequently operate under financial strain, impacting their ability to provide adequate services and maintain infrastructure.

Some examples that serve to illustrate these challenges, are the following:

- *Long Waiting Times*: A common issue reported by patients in the public sector is the lengthy wait for specialist consultations and surgeries. In some cases, patients wait months or even years for necessary medical procedures, which can lead to the worsening of their conditions.
- *Regional Disparities*: Access to healthcare services can vary significantly depending on geographic location. Rural areas, in particular, face a shortage of medical professionals and facilities, exacerbating the inequality in healthcare provision.
- *Financial Constraints*: Public hospitals often struggle with budget constraints, leading to shortages of essential medical supplies and outdated equipment. This affects the quality of care patients receive and contributes to the overall inefficiency of the system.
- *Sharp Disparity Between Private and Public Healthcare Service Management*: For anyone wanting to set an appointment with a medical specialist for a consultation or to have a medical procedure done or to schedule a surgery, where you can choose the physician and the hospital or clinic, the private healthcare system offers all of these without delay, and of course, if you have a good health insurance policy, you are pretty much covered and must only pay what the health insurance policy does not cover. For someone who is in FONASA, however, it is a much different story, and unless you have the means to complement FONASA with a private complementary health insurance, you are left on your own. Usually, a complementary health insurance takes the form of a collective coverage health insurance, provided by the employer or some other type of complementary health insurance that is individually financed. This type of insurance can help with the major costs of a serious illness or health condition, but if you do not have it, you can expect to wait a long time to get the healthcare service you need from FONASA and may very well die waiting, like many hundreds of people do every year.

2.2 Diagnosing the Illness of the Chilean Healthcare System: A Hopeful Prognosis

According to the authors Patricio Oliva Mella and Carmen Gloria Narváez in their essay entitled: “Analysis of the Chilean health system and its structure in social participation” (2022), the political and economic development that Chile has had in recent decades has allowed the evolution of its public healthcare systems to a certain extent, wherein hospital coverage, healthcare management, and infrastructure have been improved (Mella & Narváez, 2022). However, the public healthcare organization and public health policies are highly centralized and structured by the government, as pointed out by these authors, something which has allowed some positive results like the rapid implementation of several programs, including the efficient vaccination plan that Chile displayed in record time as soon as the vaccines were available in early 2021 (Mella & Narváez, 2022). However, according to these authors, there is not sufficient participation from all involved sectors and a clear lack of opportunity to voice their concerns and opinions when it comes to setting the course of public health policy. Moreover, there is permanent pressure from non-governmental groups asking for a different governance amidst a tacit rigidity of the government policy implementation and management. Therefore, there is a need for more proactiveness and more participation of the different parties involved in healthcare policy issues, although this requires a constant dialogue between the state and citizens (Mella & Narváez, 2022).

Incredibly so, despite the many challenges faced by the Chilean healthcare system, it exhibits results that position Chile favorably at the international level, even in comparison with developed countries. Thus, Chileans enjoy a longer life expectancy than the average of Organization for Economic Co-operation and Development (OECD) nations, as well as a lower mortality rate from preventable diseases. This is achieved with health expenditures that place the country within the average relative to the income level of other OECD countries (OECD, 2023). However, there is one aspect in which Chile consistently shows one of the worst performances: 61% of the population is dissatisfied with the availability of quality healthcare services. This is well above the OECD average (2023), which is 33%. Surveys indicate that this disparity highlights the main problem, which is the lack of protection felt by a large percentage of the population, where access to health benefits suffers from long waiting times in the public healthcare system FONASA and depends on the health status (pre-existences in ISAPREs) and the financial possibilities of each individual, as explained before, in terms of whether the person has or does not have access to a complementary health insurance.

2.3 The Need for a Joint Effort from the Parties Involved and a Vision for a Better Future

Despite the consensus on the need to address these challenges, progress in recent years has been limited. Despite the improvements implemented in recent decades, most notably the creation of the Explicit Health Guarantees (GES) system, that covers a number of serious illnesses and conditions, it has not been possible to establish a clear institutional framework that effectively regulates the participation of private entities in insurance or their integration into the system as a whole. Similarly, there are considerable shortcomings in both FONASA and state healthcare services, which have been documented in numerous studies (Wiff et al., 2022; Abbott, 2022).

In the meantime, the health system has faced additional strains that have exacerbated the current problem. On the one hand, the pandemic caused delays in scheduled medical care, leading to even longer waiting lists for consultations and procedures, even in those guaranteed by law. On the other hand, since the beginning of the last decade, there has been a marked increase in the judicialization of health disputes, mainly through the filing of appeals for protection, which has particularly affected ISAPREs, but is also beginning to manifest itself in other matters, such as high cost of medicines (About the Superintendence of Health, 2024). Based on the current state of affairs, efforts are now focused on the current crisis of the ISAPREs, whose financial solvency is threatened by the recent Supreme Court rulings of 2022 and 2023. The rulings would result in the generation of permanent financial imbalances and significant amounts to be returned to users for charges over several years that were illegitimate and should had never been paid by the ISAPREs' affiliates. This situation increases the risk that institutions will not meet the minimum financial solvency standards required by law, considering that in September 2023, two of them were already at the established limit (Superintendencia de Salud, 2023). In the most adverse scenario, noncompliance with these regulations would imply their closure or bankruptcy and the transfer of their affiliates to FONASA if no other ISAPRE is willing to receive them, in accordance with the Law No. 19.895 (Flores Diaz, 2021). The possible insolvency of some of these institutions would produce abrupt changes in the system that would not only affect their current users, but would also have serious repercussions on the healthcare system as a whole. Potential consequences of this crisis could include the impossibility of accessing other private insurance—especially for older and sicker people—as well as increased demand for public healthcare services with the consequent increase in waiting lists and the interruption of care by private providers, used by a significant portion of FONASA users as of today.

2.4 *Healthcare Is a Universal Right That Must Be Guaranteed for All People, Regardless of Socioeconomic Status*

Health is one of the most urgent social priorities, and there is a huge inequity in access to equal healthcare, which is a serious problem for Chile. It is complex because it touches on the essence of the human being, that is, life itself. Thus, all parties involved must make a joint effort based on pragmatism and realism to find a way out of this mess. The public and private sectors have made important contributions to the healthcare in Chile and can definitely work together to find common ground and collaborate in earnest to overcome the current crisis. A first step is to look at the healthcare system of the future where technology will play a substantive role and will change the paradigms in treating diseases. Let us think, for example, about Internet Of Things (IOT) where everything, including humans, will be connected through sensory networks, operating with 5G and 6G technologies, to optimize the great hospital challenge of connecting the patient with the hospital, medical support, and ambulance in an expeditious, timely, and efficient way (Imoize et al., 2021; Nayak & Patgiri, 2021). Likewise, the support through telemedicine in remote locations can provide effective support from the specialists or the use of nanotechnology that will allow the transfer of drugs with controlled release, obtaining greater efficacy (Adepu & Ramakrishna, 2021). Robotics is another field of medical development in which the development of artificial intelligence will allow us to have the experience not only of a surgeon but also of millions of surgeries and surgeons, obtaining a better diagnosis and/or prognosis. The development of diagnostic imaging such as magnetic resonance imaging (MRI) and computed tomography (CT) allow for early detection of diseases such as cancer of the pancreas, prostate, and other types of cancer. Technology is going to change the way we treat the diseases of the future, so we must ask ourselves, is our hospital system up to the challenge?

A high priority, in order to improve the healthcare system of Chile, should be to advance in the overhauling and modernization of the public health sector's governance, and this should also encompass its much-needed restructuring, according to the current reality and future outlook. Health spending in Chile has grown four times the average of OECD countries, yet the productivity levels in healthcare services management and practice are half that of the private healthcare sector, along with huge underutilized investments (Health at a Glance, 2023: Highlights for Chile, 2024). The last report of the National Productivity Commission (CNP) on the public healthcare system delivered to the president of the country mentions the underutilization of surgical wards by 53%, equivalent to some U$ 350 million, which could finance 115,000 surgeries (Oliveira et al., 2021). It is not necessary to build new hospitals; what is necessary is to improve the management of the current hospital infrastructure and prioritize prevention. The situation in Chile is going to be complex, according to the Chile National Socioeconomic Characterization (CASEN) survey, as it is estimated that by 2025, one in five people in Chile will be over 65 years old, and the life expectancy among men and women is currently 80.5 years, leading the region, according to WHO data (Comisión Nacional de Productividad, 2024).

The great challenge for Chile is to have a common vision where both private and public healthcare systems can coexist and can also collaborate and complement each other without having to sacrifice one in favor of the other or to have to eliminate the private healthcare alternative in favor of a public healthcare system only, where "one size fits all." There is no doubt that public hospitals and clinics must be better run, professionally managed by highly skilled people, specialized in the management and governance of healthcare, as it is in the private sector, and measured by the very same standards and metrics that the private healthcare sector uses.

Equal and timely access to healthcare services for all people in Chile is possible but requires will, courage, and stern pragmatism to undergo major reforms that can lead to efficient, quality healthcare services that amount to a social aspiration of the vast majority of people in Chile, which cannot be postponed. We must never forget that health is a social good established in the Universal Declaration of Human Rights of the United Nations: "Everyone has the right to a standard of living adequate for the health and well-being of himself/herself and of his/her family" (Universal Declaration of Human Rights of the United Nations, 2024). Thus, it should be an intrinsic part of the social sustainability principles of the country's healthcare system, which go hand in hand with the principles of social security of the population. In today's single national health insurance fund, FONASA, where approximately 80% of the population is affiliated, patients are faced with a single provider, long waiting lists, inefficiently managed public hospitals and other healthcare units such as emergency care units, and lack of resources and specialists. On the other hand, there are the ISAPREs with enough specialists ready to be consulted, enough technical and administrative resources, and sufficient and quite adequate healthcare infrastructures that serve 3.3 million Chileans with multiple plans and providers for the patient to choose from. This is where the greatest inequity in health is present, and as long as we do not fix it, people will continue to suffer.

3 Conclusions

The inequality in healthcare access in Chile is a complex issue influenced by economic, geographic, and systemic factors. The public-private divide creates a two-tiered system where access to quality care is largely determined by socioeconomic status. Geographic disparities further widen this gap, with rural populations facing significant challenges in accessing necessary services. Economic barriers and inconsistent quality of care within the public system exacerbate the issue, while bureaucratic inefficiencies and cultural factors create additional layers of complexity. Addressing these inequalities requires comprehensive reforms aimed at improving the public healthcare infrastructure, increasing funding, and ensuring equitable distribution of resources. Policies that prioritize the needs of underserved populations, streamline administrative processes, and promote cultural competence in healthcare delivery are essential for creating a more inclusive and sustainable healthcare system in Chile.

In summary, what is important for a real, comprehensive health reform that makes equal, timely, and affordable access for all Chileans a reality is an approach that modernizes our public sector by improving the management of healthcare, modernizing our public sector, and improving its governance. Efforts must be aimed at focusing spending on prevention and the primary sector, which is the axis of all public policies, giving freedom of choice of provider for FONASA patients, subsidized for those with lesser resources, a guaranteed health plan financed by the central government for low-income sectors. For catastrophic sectors of the population, those that have very little or nothing at all, a universal catastrophic insurance fund must be created wherein a public-private agreement is realized, which maximizes resources, and where a special provision in the law for health workers exists that rewards merit commitment to high standards in the practice of healthcare and to career path over greed and nepotism. We must never forget that health workers are the central axes of a comprehensive healthcare sector policy of the future.

Integrating social sustainability principles into the Chile's healthcare system involves a multifaceted approach that addresses equity, quality, affordability, and timeliness. By focusing on these principles, Chile can develop a healthcare system that not only meets the current demands of its population but also remains resilient and equitable in the long term. Social sustainability principles must also ensure that healthcare reforms are not merely reactive but proactive, fostering a healthier, more inclusive society. This holistic approach is key to transforming Chile's public healthcare system into one that delivers high-quality, affordable, and timely services to all its citizens, thereby supporting the overall social and economic well-being of the nation.

References

Abbott, T. (2022). Effect of a cost-sharing variation on the utilization of GES insurance and private healthcare services.

About the Superintendence of Health. (2024). https://www.superdesalud.gob.cl/568/w3-article-6487.html?noredirect. Accessed June 2024.

Adepu, S., & Ramakrishna, S. (2021). Controlled drug delivery systems: Current status and future directions. *Molecules, 26*(19), 5905.

Ajmal, M. M., Khan, M., Hussain, M., & Helo, P. (2018). Conceptualizing and incorporating social sustainability in the business world. *International Journal of Sustainable Development & World Ecology, 25*(4), 327–339.

Alvial, X., Rojas, A., Carrasco, R., Durán, C., & Fernández-Campusano, C. (2021). Overuse of health care in the emergency services in Chile. *International Journal of Environmental Research and Public Health, 18*(6), 3082.

Barrientos, A. (2000). Getting better after neo-liberalism: Shifts and challenges of health policy in Chile. *Healthcare Reform and Poverty in Latin America*, 94–111.

Bitran, R. (2013). *Explicit health guarantees for Chileans: The AUGE benefits package*. World Bank.

Bossert, T. J., & Villalobos Dintrans, P. (2020). Health reform in the midst of a social and political crisis in Chile, 2019-2020. *Health Systems & Reform, 6*(1), e1789031.

Charney, J., Marshall, P., & Christodoulidis, E. (2021). 'It is not 30 pesos, it is 30 years': Reflections on the Chilean crisis: Introduction. *Social & Legal Studies, 30*(4), 627–668.

Comisión Nacional de Productividad. (2024) Comisión Nacional de Productividad presenta resultados preliminares del estudio sobre el uso de los pabellones en hospitales https://www.minsal.cl/tag/comision-nacional-de-productividad/. Accessed June 2024.

Crespo, C. F. (2018). Chile: New health and institutional challenges in a country in transition. *Revista Panamericana de Salud Pública, 42*, 1.

Cristiano, S., Ulgiati, S., & Gonella, F. (2021). Systemic sustainability and resilience assessment of health systems, addressing global societal priorities: Learnings from a top nonprofit hospital in a bioclimatic building in Africa. *Renewable and Sustainable Energy Reviews, 141*, 110765.

de la Jara, J. J., & Bossert, T. (1995). Chile's health sector reform: Lessons from four reform periods. *Health Policy, 32*(1–3), 155–166.

Dominguez-Cancino, K. A., Palmieri, P. A., & Martinez-Gutierrez, M. S. (2020). National health policy reform for primary care in Chile: A qualitative analysis of the health program documents. *Journal of Primary Care & Community Health, 11*, 2150132720924884.

Flores Diaz, S. (2021). The management of the public health system through purchases of services agreements with the private sector in Chile. *Revista Digital de Derecho Administrativo, 26*, 163.

Foot, D. K., & Ross, S. (2017). Social sustainability. In *Teaching business sustainability* (pp. 107–125).

Gavurova, B., Kocisova, K., & Sopko, J. (2021). Health system efficiency in OECD countries: Dynamic network DEA approach. *Health Economics Review, 11*, 1–25.

Göttems, L. B. D., & Mollo, M. D. L. R. (2020). Neoliberalism in Latin America: Effects on health system reforms. *Revista de saude publica, 54*, 74.

Health at a Glance. (2023). *Highlights for Chile* https://www.oecd.org/chile/health-at-a-glance-Chile-EN.pdf. Accessed June 2024.

Homedes, N., & Ugalde, A. (2005). Why neoliberal health reforms have failed in Latin America. *Health Policy, 71*(1), 83–96.

Imoize, A. L., Adedeji, O., Tandiya, N., & Shetty, S. (2021). 6G enabled smart infrastructure for sustainable society: Opportunities, challenges, and research roadmap. *Sensors, 21*(5), 1709.

Letelier, A. (2014). Waiting list in a public health facility in Santiago, Chile. *Medwave, 14*(6), e6000–e6000.

Manual de la organización del Gobierno de Chile. (1959). Chile's Financial Studies Department. Chilean Budget Directorate. Published in Talleres Gráficos "La Nación'.

Manuel, A. (2002). The Chilean health system: 20 years of reforms. *Salud Pública de México, 44*(1), 60–68.

Martinez-Gutierrez, M. S., & Cuadrado, C. (2017). Health policy in the concertación era (1990–2010): Reforms the Chilean way. *Social Science & Medicine, 182*, 117–126.

Mella, P. F. O., & Narváez, C. G. (2022). Analysis of the Chilean health system and its structure in social participation. In SciELO Preprints. https://doi.org/10.1590/SciELOPreprints.3557

Nayak, S., & Patgiri, R. (2021). 6G communication technology: A vision on intelligent healthcare. In *Health informatics: A computational perspective in healthcare* (pp. 1–18).

Núñez, A., Manzano, C. A., & Chi, C. (2020). Health outcomes, utilization, and equity in Chile: An evolution from 1990 to 2015 and the effects of the last health reform. *Public Health, 178*, 38–48.

Oliveira, S. C. D., Machado, C. V., Hein, A. R. A., & Almeida, P. F. D. (2021). Public-private relations in Chile's health system: Regulation, funding and service delivery. *Ciência & Saúde Coletiva, 26*, 4529–4540.

Peña, C., & Silva, P. (Eds.). (2022). *Social revolt in Chile: Triggering factors and possible outcomes*. Routledge.

Rørdam, S. (2016). The chilean health care system.

Rotarou, E. S., & Sakellariou, D. (2017). Neoliberal reforms in health systems and the construction of long-lasting inequalities in health care: A case study from Chile. *Health Policy, 121*(5), 495–503.

Talbot, L., & Verrinder, G. (2017). *Promoting health: The primary health care approach*. Elsevier Health Sciences.

Universal Declaration of Human Rights of the United Nations. (2024). https://www.un.org/en/about-us/universal-declaration-of-human-rights. Accessed June 2024.

Villalobos Dintrans, P. (2020). Health systems, aging, and inequity: An example from Chile. *International Journal of Environmental Research and Public Health, 17*(18), 6546.

Wiff, J. I., Benítez, A. F., Pezoa, M., González, C. P., Chepo, M., & Flores, R. L. (2022). The National Health Policy for International Migrants in Chile, 2014–17. International Development Policy, 338.

World Health Organization. (2023). *National health policies, strategies and plans*. Available online: http://www.who.int/nationalpolicies/about/en/. Accessed on 27 June 2023.

Zurynski, Y., Herkes-Deane, J., Holt, J., McPherson, E., Lamprell, G., Dammery, G., et al. (2022). How can the healthcare system deliver sustainable performance? A scoping review. *BMJ Open, 12*(5), e059207.

Zurynski, Y., Ludlow, K., Testa, L., Augustsson, H., Herkes-Deane, J., Hutchinson, K., et al. (2023). Built to last? Barriers and facilitators of healthcare program sustainability: A systematic integrative review. *Implementation Science, 18*(1), 62.

The Impact of Sustainable Technologies on Business Strategy and Competitiveness

Emilio Abad-Segura, Francisco José Castillo-Díaz, Ana Batlles-delaFuente, and Luis J. Belmonte-Ureña

Contents

1 Introduction

The global disruption caused by the COVID-19 pandemic has compelled businesses to undergo digital transformation, reshaping their models and responding to evolving customer needs (Li et al., 2020). This demand for innovation is fueled by the

E. Abad-Segura · F. J. Castillo-Díaz (✉) · A. Batlles-delaFuente · L. J. Belmonte-Ureña
Department of Economy and Business, University of Almeria, Almeria, Spain
e-mail: eas297@ual.es; fcastillo@ual.es; anabatlles@ual.es; lbelmont@ual.es

S. S. Muthu (ed.), *Assessment of Social Sustainability Management in Various Sectors*, Environmental Footprints and Eco-design of Products and Processes,
https://doi.org/10.1007/978-3-031-81029-9_6

recognition that technological innovation (TI) is vital for economic growth. As a result, technology-driven disruptions are redefining business processes, propelled by disintermediation within the value chain through Internet-based business models and mobile ecosystems. These transformations enhance business competitiveness and operational efficiency (Healy et al., 2018), while allowing for a better contextual analysis of business operations and streamlining of accounting systems.

The integration of cloud-based digital infrastructure and advancements in information technologies has automated critical business operations, offering a nuanced contextual analysis of financial transactions. Technologies such as artificial Intelligence (AI), machine learning (ML), big data, 5G, virtual reality, augmented reality, blockchain, cloud computing, robotic process automation (RPA), and the Internet of Things (IoT) directly impact organizations, enabling strategic planning and informed decision-making (Desplebin et al., 2021; Mancini et al., 2021).

As companies navigate changing customer, worker, and societal behaviors, the imperative arises for them to evolve into digital entities, fostering innovative digital transformation strategies across customer relationships, channels, platforms, organization, and human capital. However, uncertainties and risks pertaining to technology, regulation, data protection, cybersecurity, and interoperability necessitate a judicious alignment between technology and business to drive innovation and value generation (Stonehouse & Snowdon, 2007; Tikhonov & Zelentsova, 2021).

Within this context, sustainable technological innovation (STI) in business involves adopting technologies and practices that balance economic profitability with minimizing environmental impacts (Urraca-Ruiz & Laguna-Molina, 2016; Williams, 2005). These include investments in cleaner technologies, such as renewable energy and fleet electrification, which align with stringent regulations and foster positive environmental impacts (Awawdeh et al., 2022; Huo et al., 2022). The subsequent sections of this chapter are structured to explore the impact of STI on enhancing business operations and competitiveness. Firstly, the theoretical framework of this research is presented. Secondly, the methodology used to carry out this work is presented. Thirdly, the results of this work are presented. Fourthly, the discussion of the results obtained is presented. Finally, fifthly, the conclusions obtained in this book chapter are presented.

2 Framework Overview

2.1 Technological Innovation in Business

The digital revolution, as emphasized by Tom Goodwin, transcends mere technology adoption; it signifies a break from outdated processes and strategies (Pavlatos & Kostakis, 2018). Joseph Schumpeter's distinction between invention and innovation, dating back to 1934, underscores the practical application of new ideas, positioning innovation as central to economic discourse (Witt, 2002). Innovation, as a response to contemporary economic challenges, faces a global issue: a limited number of companies basing competitiveness on innovation (Bartolacci et al., 2015).

Technological innovation (TI) goes beyond acquiring technology; it involves process renewal to interact with technology effectively and maximize performance. TI encompasses four types: incremental, disruptive, sustainable, and radical (Dai & Vasarhelyi, 2017). The term "technological innovation" is preferred in academic discourse, signifying a focus on improving business value through technological aspects rather than working on technology for its own sake. The convergence of different technologies contributes to product or service success.

TI mandates a shift in mentality, necessitating qualified professionals. Companies, compelled to adapt to market changes, rely on TI, extending beyond research and development (R&D). It involves all phases essential for successful product development, commercialization, and the introduction of new services (Dai & Vasarhelyi, 2017).

TI aligns with digital Darwinism, updating Charles Darwin's theory to elucidate how technology evolution transforms society and business management, impacting production processes, customer communication, and sales (Wonglimpiyarat, 2016). Automation of processes through new technologies enhances a company's administrative functions, ensuring data quality, report agility, professional diversification, increased synergy, diverse approaches, and higher productivity (Gardner & Bryson, 2021; Healy et al., 2018).

The Fourth Industrial Revolution (4IR or Industry 4.0) assigns a transformative role to innovative technologies, influencing production models, labor, and consumer relations. Data-driven technology trends reshape accounting and finance disciplines, contributing to a global shift toward digital transformation (Babenko et al., 2021).

TI facilitates greater efficiency, flexibility in process management, incorporation of automation and optimization methods, implementation of business intelligence systems, and job creation and emphasizes crucial skills for companies, such as critical thinking and flexibility (Akter et al., 2020; Liu & Stephens, 2019). AI solutions, including robotic process automation (RPA) and machine learning (ML), have revolutionized industries, automating tasks without job replacement. AI aids professionals by focusing on impactful actions and value creation (Marushchak et al., 2021; Petkov, 2020).

Technology utilization correlates with a company's ability to enhance growth within a specific socioeconomic environment, emphasizing business competitiveness. Factors influencing competitiveness encompass management capacity, quality-price relationship, employee experience, talent acquisition, community building with clients, internationalization, technological resource selection, connectivity, and research and development expenditure (Ketels, 2006). In this sense, in addition to costs, there are other factors that will affect the competitiveness of a company, such as management capacity, the relationship between quality and price, employee experience and satisfaction, attracting and retaining talent, building community with the client, the internationalization of the company, the ability to select the necessary technological resources according to their scope, the technology that allows connectivity, and spending on R&D.

Innovation, a fundamental factor for development in advanced countries, goes beyond technology incorporation; it anticipates market needs and detects new products, processes, and services, fostering new features at minimal cost. Innovative strategies, organizational environmental changes, and a commitment to sustainable technological innovation (STI) are imperative for business competitiveness (Stonehouse & Snowdon, 2007). Large companies play a crucial role in fostering innovation, which has become increasingly vital due to global competition, shorter product life cycles, and evolving consumer demand (Gardner & Bryson, 2021).

Planning thus offers the ability to enhance international competitiveness, as customers are free to accept international offers due to lower prices, compelling national producers to adapt their strategies accordingly. From this perspective, business competitiveness can be understood in three forms: structural, which relates to the quality of a company's internal operating system or structure; economic, which ties to the ability to produce or offer products or services at a competitive price and without losses or risks that jeopardize the business's longevity; and systemic, which positions it within its economic environment and addresses competitiveness from the level of relationships between a group of companies and other economic actors (Sisaye & Birnberg, 2014; Tikhonov & Zelentsova, 2021).

In this context, a flexible technological platform becomes a fundamental tool for business competitiveness. Such platforms enable the integration of knowledge about the client, debtor, creditor, and supplier, along with all their interconnections. This capability facilitates the articulation of more efficient models of production and distribution of services, which, in addition to opening new markets, enable the economic development of countries and the inclusion of disadvantaged sectors (Stonehouse & Snowdon, 2007). Furthermore, there is a noted relationship between import activities and TI and non-TI because it encourages the introduction of product, process, marketing, and organizational innovations in companies from emerging markets. Similarly, in the context of business disruptions stemming from the COVID-19 pandemic, there are significant economic challenges for companies in emerging markets. Consequently, some governments have adopted protectionist policies, which can create barriers against imports and hinder the innovative success of local organizations (Montégu et al., 2021).

2.2 *Sustainable Technological Innovation to Enhance Business Competitiveness*

Sustainable technological innovation (STI), a paramount concept in contemporary business, harmonizes technological progress with environmental preservation and the well-being of future generations. This approach gains significance in addressing environmental challenges like climate change, biodiversity loss, and resource scarcity (Pavlatos & Kostakis, 2018). This approach is increasingly significant as businesses face environmental challenges like climate change, biodiversity loss, and resource scarcity.

In the current business landscape, STI emerges as a pivotal strategy for increasing competitiveness, aligning with the growing importance of environmental awareness and social responsibility. This innovative paradigm aims for excellence in efficiency and productivity, integrating environmental and societal impacts into its considerations. Companies, in pursuit of enhanced competitiveness, adopt approaches from energy efficiency to circular economy (CE), digitalization, and sustainable practices across the value chain (Dai & Vasarhelyi, 2017).

A cornerstone of STI lies in solutions promoting energy efficiency and reducing greenhouse gas emissions. This entails developing clean, renewable technologies such as solar, wind, and hydroelectric energy to mitigate climate change. Energy efficiency improvements span various sectors through smart devices and systems, minimizing energy waste (Bartolacci et al., 2015). Sustainable management of natural resources is a crucial aspect of STI, incorporating practices like precision agriculture and promoting CE to reduce, reuse, and recycle products and materials (Dai & Vasarhelyi, 2017).

Research underscores energy efficiency as a key tenet of STI embraced by companies. Implementing technologies like energy-efficient LED lighting and intelligent energy management systems reduces operational costs and carbon footprints. Striking a balance between cost reduction and environmental impact mitigation enhances competitiveness in a market where sustainability is increasingly valued (Petkov, 2020).

Companies, in their pursuit of heightened competitiveness, are shifting toward sustainable energy sources, reducing reliance on fossil fuels, and showcasing commitment to a cleaner future. This transition not only mitigates vulnerability to price fluctuations but also signifies dedication to sustainability, potentially resulting in significant long-term cost savings (Babenko et al., 2021).

Digitalization and automation, integral to STI, employ technologies like the Internet of Things (IoT), artificial intelligence (AI), and data analytics. These technologies optimize operations, enhance decision-making, and offer more efficient, personalized products and services. Digitalization increases productivity, reduces resource consumption, and minimizes waste, bolstering cost competitiveness and positioning companies as innovative industry leaders (Palmié et al., 2020).

Circular economy (CE), another key component of STI, contrasts with the linear model of production and consumption. It encourages companies to reduce, reuse, and recycle materials, employing strategies like efficient waste management, designing products for longer lifespans, and promoting material reuse. CE not only lessens the environmental impact but also fosters innovation in products and services, offering new business opportunities and increasing competitiveness (Bröring et al., 2020; Watari et al., 2019). Therefore, CE reduces the negative environmental impact of business operations, in addition to generating new business opportunities and increasing competitiveness through innovation in products and services. In this sense, several research streams consider that CE is another essential component of STI. Instead of following a linear model of production and consumption, companies are adopting a circular approach that seeks to reduce, reuse, and recycle materials and resources (Morales et al., 2021). Implementing more efficient waste management systems, designing products with a longer useful life, and promoting the reuse of products and materials are key strategies in this regard (Abad-Segura et al., 2020). CE not only decreases the negative environmental impact of business operations, but can also generate new business opportunities and increase competitiveness through innovation in products and services.

Collaboration among governments, businesses, and civil society is pivotal in advancing STI. Policies and regulations incentivizing the adoption of clean and sustainable technologies are crucial for a swift transition to a more sustainable future. Businesses contribute by investing in sustainable technology research and development while adopting responsible practices (Dai & Vasarhelyi, 2017).

3 Material and Methods

To explore the scientific insights into the nexus between sustainable technological innovation (STI) and business competitiveness, we conducted a robust quantitative analysis using statistical techniques. This was complemented by a qualitative systematic review based on a sample from the Scopus database, which is known for its extensive coverage and reliability in scholarly literature (Abad-Segura & González-Zamar, 2020; Gureyev & Mazov, 2022; Hota et al., 2020). Despite inherent methodological limitations and potential biases introduced by the database selection, Scopus ensures comprehensiveness by encompassing articles found in other databases like Web of Science (WoS) or Google Scholar.

In adherence to the Preferred Reporting Items for Systematic Reviews and Meta-analyses (PRISMA) method (Moher et al., 2011), we meticulously identified 533 articles through systematic sampling. Employing processing and mapping tools, including the VOSviewer software (version 1.6.19, Leiden University, Leiden, Netherlands), allowed for the identification of research clusters and trends over time (Ding & Yang, 2020; Van Eck & Waltman, 2010; Waltman et al., 2010).

The modularity-based VOSviewer clustering approach, which integrates mapping and clustering methods, facilitated a comprehensive analysis of keyword relevance (Waltman & van Eck, 2012). This scoring system aids in predicting future research trajectories. Additionally, a detailed examination of the evolution of scientific production over different periods and thematic areas was conducted using bibliometric indicators (Fig. 1, Table 1).

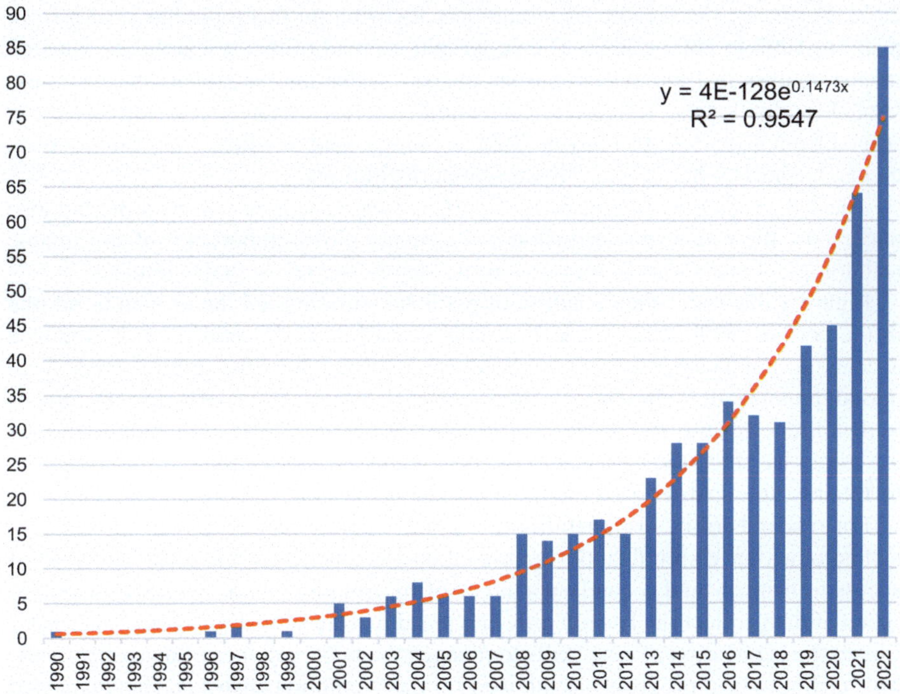

Fig. 1 Evolution and trend line of the scientific literature (1990–2022)

Table 1 Evolution of the scientific literature by subperiods (1990–2022)

Subperiod	Articles	Journals	Authors	Countries	Institutions	Citations
2020–2022	194	118	195	67	463	1719
2010–2019	265	134	261	64	488	4842
2000–2009	69	48	65	37	111	418
1990–1999	5	4	10	25	10	138

4 Outcomes

4.1 Evolution of Scientific Production

Table 1 shows the evolution of the most outstanding characteristics of the samples of articles published during the period 1990–2022, by decades, except for the first period of 3 years. The number of published documents increased from 1 in 1990 to 85 in 2022. During the last 5 years (2018–2022), 267 articles (50.09%) were published. If we extend the period to the last decade (2013–2022), the accumulated publications were 412 (70.30%).

The number of journals featuring these contributions has also expanded dramatically, from only the *Harvard Business Review* in 1990 to 56 different journals in

2022. Similarly, the breadth of researchers focusing on this topic has grown from just 3 in 1990 to 160 in 2022. The geographic spread of this research has widened from 1 country in 1990 to 91 countries in 2022, reflecting the global interest in sustainable technologies. Additionally, the number of research institutions involved has surged from 1 in 1990 to 1064 in 2022, encompassing faculties, university departments, business schools, and other research institutions.

These data not only highlight the significant increase in research output but also underscore the widespread diffusion and growing global importance of sustainable technological innovation across various academic and research disciplines.

Figure 1 illustrates the evolution of published articles and the growth trend line in the period 1990–2022. The exponential trend line aptly signifies a pronounced acceleration in the quantity of published articles throughout the scrutinized 33-year period. Notably, the robust statistical metric of determination coefficient (R^2) attests to the efficacy of this trend line, with a value of 0.9547. This coefficient delineates the extent to which the variance in the dependent variable (number of articles) can be reliably anticipated based on the independent variable (year of publication).

For researchers in sustainability and related fields, innovations in information technologies have significantly facilitated advancements in this area. The ease of access to primary and secondary sources of information, along with the increased speed of data collection, has not only enhanced research capabilities but also necessitated the acquisition of new skills to manage the large volumes of data effectively. This technological progress has allowed for more comprehensive and sophisticated analyses of trends and impacts in the realm of sustainable technological innovation (Abernethy & Bouwens, 2005; Goodman & Palmon, 2001).

The search was limited to the thematic areas of "business, management and accounting," which included 430 articles (80.68%), and "economics, econometrics, and finance" with 243 (45.59%). An article can be classified in both thematic areas, depending on which is linked to the journal where it is published.

The "business, management, and accounting" thematic area is aimed at developing, the topics, in the following order: sustainability, sustainable development, technology transfer, Industry 4.0, entrepreneurship, sustainable innovation, green innovation, product innovation, knowledge management, economic growth, eco-innovation, and dynamic capabilities. The search was limited to the thematic areas of "business, management, and accounting," which included 430 articles (80.68%), and "economics, econometrics, and finance" with 243 (45.59%). An article can be classified in both thematic areas, depending on which is linked to the journal where it is published.

The "business, management, and accounting" thematic area is aimed at developing, the topics, in the following order: sustainability, sustainable development, technology transfer, Industry 4.0, entrepreneurship, sustainable innovation, green innovation, product innovation, knowledge management, economic growth, eco-innovation, and dynamic capabilities.

Meanwhile, the subject area "economics, econometrics, and finance" focuses mainly on sustainability, sustainable development, environmental economics, technological development, technology transfer, sustainable growth, economic development, competitiveness, industrial performance, globalization, and energy efficiency. This illustrates a strong emphasis on how economic theories and practices are adapting to the challenges and opportunities presented by sustainable and technological advancements.

Within the context of our globalized world, interdisciplinary research has emerged organically as a response to the progressive expansion of knowledge necessitating a heightened comprehension of anthropological, ethnic, linguistic, and sociocultural diversity. This approach represents a paradigm shift, where various disciplines are integrated into a cohesive analytical framework, enhancing the depth and breadth of analysis. Such integration is crucial for addressing complex global issues that span multiple sectors and disciplines (K. C. Chan et al., 2013; Islam & Agarwal, 2020).

4.2 Analysis of Knowledge Structure

To elucidate the evolution of research trends, a meticulous analysis of the keywords employed in the 533 articles was conducted. Table 2 shows the 40 most used keywords, which represent the trends during the period 1990–2022. From the search of the descriptors, 2056 different keywords have been identified.

Without considering the search descriptors, the most used term was "sustainable development," reference of cluster 3, which refers to the search for a balance between economic growth, social equity, and environmental preservation, guaranteeing the needs of present and future generations. Among these terms, 45% are linked to cluster 1 (competitiveness), among others, with technological development, dynamic capabilities, sustainable innovation, small and medium enterprises (SME), or R&D. It is then followed by clusters 3 (green Innovation, 25%), 2 (Industry 4.0, 17.50%), 5 (supply chain management, 10%), and 6 (environmental economics, 2.5%).

Table 3 shows the three main keywords used in each decade. In the first period (1990–1999), research contributed to a deeper understanding of organizational innovation. These advances meant greater collaboration between companies and entrepreneurs, accelerating the adoption of innovations and promoting a more dynamic and competitive business environment. This integrated approach has proven essential for economic prosperity and business transformation in the contemporary era.

In the subsequent decade (2000–2009), progress was made on how the principles of sustainable development can drive technological innovation. Research has shown that policies and regulations that promote sustainability can stimulate investment in research and development of clean and socially responsible technologies. In addition, it explored how consumer demand and investor pressure for sustainable practices drive innovation in products and services that are more respectful of the environment and society.

Table 2 Most frequently used keywords

Rank	Keyword	Cluster	Articles	%	Rank	Keyword	Cluster	Articles	%
1	Sustainable development	3	73	13.70	21	Innovation management	3	6	1.13
2	Sustainability	2	66	12.38	22	Supply chain management	5	6	1.13
3	Industry 4.0	2	11	2.06	23	Competitive advantage	1	5	0.94
4	Competitiveness	1	10	1.88	24	Globalization	1	5	0.94
5	Technological development	1	10	1.88	25	Open innovation	1	5	0.94
6	Circular economy	2	10	1.88	26	Sustainable Development goals	1	5	0.94
7	Green innovation	3	10	1.88	27	Sustainable entrepreneurship	1	5	0.94
8	Dynamic capabilities	1	9	1.69	28	Innovation policy	3	5	0.94
9	Sustainable innovation	1	9	1.69	29	Patents	3	5	0.94
10	Eco-innovation	3	9	1.69	30	Product development	3	5	0.94
11	Environmental economics	6	9	1.69	31	Consumption behavior	5	5	0.94
12	SME	1	8	1.50	32	Digitalization	5	5	0.94
13	Product innovation	3	8	1.50	33	Climate change	1	4	0.75
14	Business model innovation	2	7	1.31	34	Eco-efficiency	1	4	0.75
15	Business models	2	7	1.31	35	Innovation process	1	4	0.75
16	Knowledge management	3	7	1.31	36	Social entrepreneurship	1	4	0.75
17	Service innovation	3	7	1.31	37	Sustainable competitive advantage	1	4	0.75
18	Technology adoption	5	7	1.31	38	Sustainable growth	1	4	0.75
19	R&D	1	6	1.13	39	Technological change	1	4	0.75
20	Digital transformation	2	6	1.13	40	Fintech	2	4	0.75

Table 3 Most used keywords by period (1990–2022)

Period	Keyword 1	Keyword 2	Keyword 3
2020–2022	Sustainability	Sustainable development	Industry 4.0
2010–2019	Sustainable development	Sustainability	Entrepreneurship
2000–2009	Sustainable development	Entrepreneurship	Product development
1990–1999	Organizational innovation	Entrepreneurship	Diffusion of innovation

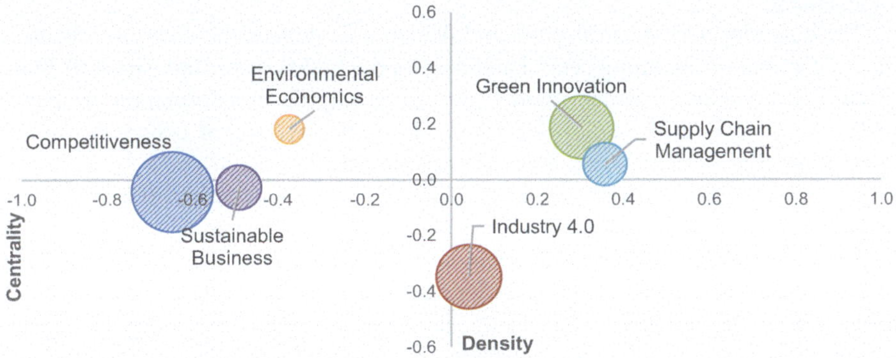

Fig. 2 Cluster centrality-density diagram. Own elaboration

In the subperiod (2010–2019), the analysis is strengthened with more research on the approach of the previous decade, and the research is linked to the guide to facilitate the implementation of the Sustainable Development Goals (SDGs), allowing governments, organizations, and civil society to take informed measures to achieve sustainable development through fostering the partnership between IT and corporate social responsibility (CSR) globally.

Lastly, in the 2020–2022 period, the publications theoretically and practically confirm the application of sustainable systems of innovative technologies, in addition to how IT drives the transformation towards Industry 4.0, which has led to improved efficiency, the creation of new business models, and workforce adaptation in an ever-evolving business environment.

4.3 Strategic Matrix

Figure 2 shows a classification of the six research lines detected based on the criteria of density and centrality within the strategic matrix. Centrality indicates the most central, most powerful, and prestigious thematic axes, that is, it measures the degree of the dominant cluster in the network; meanwhile, density is considered a measure of cohesion between the elements of the network, so the higher the density, the greater the cohesion and the greater the number of links between the elements that make up the network (Li et al., 2021).

The quadrant with centrality (+) and density (+) includes the central themes and with a high degree of development: "green innovation" and "supply chain management." The first is linked to terms like eco-innovation, product innovation, innovation management, innovation policy, patents, environmental innovation, radical innovation, biotechnology, efficiency, or employment; meanwhile, the second does it with digitalization, e-commerce, green economy, corporate strategy, retailing, foreign direct investment (FDI), Global Innovation Index (GII), or stakeholder engagement.

The quadrant with centrality (+) and density (−) defines the emerging themes, that is, capable of becoming central and developed. In this case, "Industry 4.0" links to circular economy, business model innovation, digital transformation, financial inclusion, fintech, ICT, social innovation, bioeconomy, digital economy, digital innovation, digital technology, or disruptive innovation.

Hence, the quadrant with centrality (−) and density (−) defines the peripheral and specialized lines of research.

This includes the "sustainable business" cluster, which has focused on the study of technology management, cooperation, diffusion of innovation, environmental protection, environmental sustainability, industrial development, renewable energy, energy policy, environmental investment, environmental technology, integration, IoT, or invention. It also includes the "competitiveness" cluster, which has developed terms such as human capital, dynamic capabilities, marketing, sustainable innovation, SME, R&D, strategy, competitive advantage, globalization, open innovation, Sustainable Development Goals, sustainable entrepreneurship, climate change, eco-efficiency, financial development, or innovation process.

The quadrant with centrality (−) and density (+) has peripheral and underdeveloped themes, being here the "environmental economics" cluster that have developed aspects such as carbon emission, energy efficiency, environmental policy, grassroots innovation, green finance, international competitiveness, entrepreneurship education, technology transfer offices (TTOS), third stream of income, university-business cooperation (UBC), university-industry innovation networks (UIINS), or co-productive institutions.

In the context of an increasingly globalized world, the imperative for interdisciplinary research has grown significantly, as noted by Chan et al. (2013) and Islam & Agarwal (2020), and has become imperative. This approach is not merely beneficial but essential, as it enhances the richness and depth of analysis by incorporating diverse perspectives from fields such as anthropology, ethnicity, linguistics, and sociocultural studies. Such integration broadens the scope of sustainable technological innovation, highlighting its multifaceted impacts across various dimensions of society.

5 Discussion

Sustainable technological innovation (STI) acts as a multifaceted strategy crucial for addressing environmental, social, and economic challenges in a cohesive and sustainable manner. By integrating advanced technologies across various domains, STI paves the way for a cleaner, more resilient, and prosperous future for both current and future generations. This approach is indispensable not only for addressing global issues like climate change and biodiversity loss but also for unlocking new avenues for economic growth and enhancing quality of life worldwide.

The evolution of STI into a robust academic discipline necessitates the application of refined scientific methods to propel societal progress and deepen understanding. In today's business environment, organizations increasingly depend on information and communication technologies (ICT) to optimize processes and enhance decision-making. This transition, while beneficial, introduces challenges that require automation to maintain competitiveness and necessitates agile decision-making in operations like accounting (Healy et al., 2018; Sisaye & Birnberg, 2014).

In a globalized context, generating new knowledge and translating it into innovative products and services becomes imperative for sustaining and enhancing economic competitiveness. In particular, developed economies require a concentrated accumulation of knowledge in service sectors and industries to foster economic growth and innovation. Universities and academic institutions play a crucial role in advancing STI, with collaboration networks being instrumental in ensuring access to high-quality technical and human resources. These networks drive research, development, and innovation while facilitating the dissemination and transmission of new knowledge (Abernethy & Bouwens, 2005).

5.1 *Schools of Thought (1990–2022)*

This analysis categorizes research into six thematic clusters, each delineating evolving research trends in STI. These clusters include "Industry 4.0," "competitiveness," "green innovation," "environmental economics," "supply chain management," and "sustainable business." Each theme underscores the interdisciplinary nature of STI and its extensive impact across different sectors.

Industry 4.0

Research asserts that the Fourth Industrial Revolution (4IR) signifies a revolutionary technological leap in manufacturing and industry, intertwining with STI to confront contemporary and future societal challenges. 4IR assimilates advanced technologies, including automation, AI, IoT, robotics, and data analytics, fostering sustainability through enhanced energy efficiency. The integration of digital industrial technology contributes to sustainability by optimizing processes, ensuring efficient energy utilization, and diminishing the carbon footprint in industrial operations.

The progression of automation and information technology since the mid-1990s has fundamentally reshaped industrial production, placing a strong emphasis on efficiency and productivity. The concept of "Industry 4.0" gained prominence in the early 2010s, marking a shift toward the convergence of technologies such as IoT, AI, machine learning, and cloud computing to revolutionize manufacturing. This shift moved the focus from merely automating processes to creating smart factories capable of continually adapting and optimizing operations. Connectivity and data collection have become essential components of 4IR, allowing companies to leverage IoT to connect machines, sensors, and production systems in real time, enabling monitoring and control from any location (Hopkins, 2021).

The collection and analysis of big data have become crucial to improve operational efficiency, predict machinery failures, and optimize supply chain management (SCM). AI and machine learning enable machines to make informed decisions and learn from experience, enhancing production optimization, anomaly detection, predictive maintenance, and product customization. Autonomous control systems have become a reality, facilitating smaller batch production and rapid adaptation to changing market demands.

Moreover, cybersecurity has emerged as a critical concern in 4IR. As more devices and systems connect to the network, research into protecting critical data and systems from potential cyberattacks has intensified. IoT security and protection against cyber threats have become essential research fields to ensure data integrity and confidentiality in manufacturing (Hoosain et al., 2020).

4IR has not only revolutionized the manufacturing sector but has also transformed supply chain dynamics, enhancing visibility and collaboration among stakeholders. Research has explored the application of 4IR technologies in inventory management, logistics, and on-demand production, underscoring the indispensability of an efficient supply chain in the interconnected global landscape. Sustainability has become a central focus, exploring energy efficiency, waste reduction, and sustainable production methods. Recent studies have investigated how companies can adapt to emerging challenges and bolster the resilience of their operations in the face of uncertainties.

Competitiveness

The thematic axis of competitiveness scrutinizes the economic aspects of globalization, underlining the necessity for companies to devise novel strategies to remain competitive. Recent decades have witnessed a paradigm shift, with increased focus on competitiveness intertwined with information technology (IT), aiming to reduce processes that incur unnecessary costs. The intellectual knowledge of organizations, encompassing intangible assets such as knowledge, skills, and organizational structures, plays a key role in enhancing competitiveness (Christensen & Nikolaev, 2013).

This line examines the intellectual knowledge of an organization referred to as the combination of intangible assets, which allows it to function and, in turn, increases its value (Ousama et al., 2011; Wyatt, 2001). Studies have found that this concept is made up of knowledge, skills, experience, information systems, intellectual property, organizational structures, etc., which can be used to create wealth

(Chenhall & Langfield-Smith, 1999; Klein, 2009). Thus, managerial management focuses on understanding and controlling costs, aiding in decision-making, and facilitating strategic planning. Numerous studies highlight that accounting management is essential for the success of a company as it provides efficient control of accounts and finances, serving as an instrument that facilitates the control process through budget preparation and the analysis of deviations (Comiskey & Groves, 1971; Sisaye & Birnberg, 2014).

Research has also concluded that transformational leadership and organizational learning significantly influence strategic management accounting, positively impacting company performance (Phornlaphatrachakorn, 2019). As technology advances, research in competitiveness has shifted toward digitalization, AI, and IoT. This shift examines how these cutting-edge technologies can drive efficiency and sustainability within corporate structures.

Moreover, the necessity for collaboration between diverse entities such as companies, governments, nonprofit organizations, and civil society has become increasingly evident. This collaboration is crucial for addressing complex environmental and social challenges effectively. In this context, public-private partnerships, business alliances, and co-creation strategies have emerged as key strategies in fostering competitiveness. These collaborative efforts help pool resources, expertise, and capabilities, creating synergies that enhance innovative solutions and market adaptability (Tohănean et al., 2020).

In conclusion, the "competitiveness" theme underscores the essential integration of technological innovation with strategic business practices to maintain a competitive edge in a rapidly evolving global market. This approach not only facilitates operational efficiencies but also supports sustainable practices that are increasingly important in today's business environment.

Green Innovation

The research trajectory in green innovation has undergone a transformative evolution, progressing from a focus on clean technologies to a holistic interdisciplinary perspective. Initially concentrated on the development of clean technologies, research in the 1990s and 2000s aimed at understanding and enhancing the efficacy of these technologies in reducing environmental impact. As sustainability gained importance, the scope of research broadened, encompassing economic and social dimensions. Interdisciplinary studies explored political, economic, and cultural barriers hindering the adoption of sustainable practices.

Environmental economics, particularly in the 1990s, revolved around the internalization of environmental costs, incorporating adverse environmental effects into economic calculations and governmental policies. Market instruments such as carbon taxes and emissions trading systems were investigated to address climate change and pollution. Over the years, environmental economics has embraced a more comprehensive approach, considering planetary boundaries and long-term sustainability. It has delved into ecological economics and complex systems theory, examining the intricate relationship between the economy and the biosphere.

Circular economy (CE) emerged as a central topic in environmental economics, emphasizing resource efficiency and waste reduction (Smith, 2016).

With advancements in technology and data collection capabilities, environmental economics has effectively utilized big data and AI to better understand and tackle environmental challenges. Research has expanded into the application of these tools for water management, natural disaster prediction, biodiversity conservation, and crafting informed environmental policy. This integration of technology facilitates a more nuanced approach to environmental management, ensuring that sustainability initiatives are based on robust, data-driven insights.

Environmental economics has also gained prominence in shaping governmental policies, providing frameworks to effectively address pressing issues like climate change, natural resource management, and environmental conservation. As sustainability concerns have moved to the forefront, research has increasingly focused on integrating sustainable investments and human well-being measurements into traditional economic models. The recent focus includes examining the interconnections between human health, the economy, and environmental health, exploring how health crises impact sustainability and how recovery policies can be designed with sustainability considerations in mind (de Jesus & Mendonça, 2018).

Environmental Economics: A Holistic Perspective

Environmental economics scrutinizes the drivers of economic growth across regions, categorizing them into two main sets of variables. The first set encompasses economic factors such as innovation, openness to the external world, productive structures, and entrepreneurial capabilities. The second set delves into business-related variables, including ownership structures, the utilization of accounting information for decision-making, and CSR policies (Englmann, 1994; Mishra et al., 2020). Grounded in neoclassical theory, economic growth is traditionally assessed through the factors of production—labor and capital. This perspective remains key in estimating the pace of technological progress (Jamali et al., 2011).

Over the last three decades, heightened awareness regarding environmental issues has prompted the integration of environmental considerations into economic and political analyses. In the 1990s, environmental economics primarily focused on internalizing environmental costs, incorporating adverse environmental effects into economic calculations, and shaping government policies. Significant research efforts centered on designing market instruments like carbon taxes and emissions trading systems to address climate change, air pollution, and water pollution (Batlles-delafuente et al., 2021).

The progression into the 2000s witnessed a paradigm shift, with environmental economics adopting a more comprehensive approach that considers planetary boundaries and long-term sustainability. Researchers explored interdisciplinary topics such as ecological economics and complex systems theory, aiming for a holistic understanding of the relationship between the economy and the biosphere. Economic models incorporating environmental variables, including biodiversity loss and land degradation, were developed to evaluate the impact of economic policies and practices on ecosystem health. Circular economy (CE) emerged as a central theme, recognizing the importance of waste reduction, resource efficiency, and recycling in sustainable resource management (Kar et al., 2016).

Advancements in technology and extensive data collection have empowered environmental economics with novel tools and approaches. Research now leverages big data and AI to comprehend and address environmental challenges, spanning water management, natural disaster prediction, and biodiversity conservation. Environmental economics has become more integrated into government policy and decision-making processes. Governments and international organizations have embraced approaches based on environmental economics to address complex issues like climate change, natural resource management, and environmental conservation. A focus has emerged on evaluating the effectiveness of environmental policies and regulations and exploring avenues for improvement to achieve more sustainable outcomes.

Currently, the research on environmental economics extends its lens to how economic policies and practices can contribute to achieving SDGs. Studies explore the incorporation of sustainable investments, CSR practices, and the measurement of human well-being into traditional economic models. Furthermore, there is a growing emphasis on the interconnectedness between human health, the economy, and environmental health. Research is increasingly focused on understanding how health crises influence sustainability and on designing recovery policies that align with sustainability principles.

This holistic perspective underscores the vital role of environmental economics in shaping a sustainable and resilient future. By aligning economic strategies with environmental sustainability, the field continues to influence global economic policies and practices, ensuring that growth and development proceed without compromising the planet's health and future.

Transformational Trends in Supply Chain Management

In recent decades, supply chain management (SCM) research has undergone dynamic evolution, reshaping companies' approaches to this pivotal operational facet. During the 1990s, SCM primarily fixated on operational efficiency and cost reduction. Research emphasis centered on inventory optimization, demand management, and augmenting the efficiency of logistics operations. A vision emerged of a more agile and efficient supply chain as a cornerstone for gaining a competitive advantage.

The early 2000s marked a transformative period as globalization and the surge in information technology revolutionized SCM research. The focus shifted toward supply chain integration and visibility, propelled by advancements in communication and collaboration. Research explored how companies could exploit information and ICT to establish transparent and interconnected supply chains, fostering superior coordination among suppliers, manufacturers, and distributors. The critical examination of supply chain risk management gained prominence, particularly following disruptive events such as September 11, 2001 and the 2008 financial crisis.

The rise of e-commerce and heightened consumer demand for expedited and flexible delivery options prompted a heightened focus on logistics and transportation. Research delved into last-mile optimization, same-day and next-day delivery, as well as fleet management and urban logistics. Notably, innovation within SCM gained attention with sustainability at its core, highlighting strategies to reduce the carbon footprint of supply chains through sustainable transport and waste management (Plaza-Úbeda et al., 2020).

The 2010s witnessed the ascension of data analytics and AI as formidable tools in SCM. Research explored how companies could leverage big data and advanced algorithms to make more informed decisions regarding inventory, demand forecasting, and supply chain planning. SCM evolved into a more demand-driven and personalized domain as companies adapted to individual customer preferences. Concurrently, the landscape of supply chain risk management expanded to include cyber threats and challenges related to data security.

Recent years have emphasized the importance of environmental and social impacts within supply chains, driven by consumer expectations and regulatory demands. Research has increasingly focused on sustainable transport methods and waste management to reduce supply chains' environmental footprints. The COVID-19 pandemic highlighted the need for supply chain resilience, accelerating research into how global supply networks can better withstand such disruptions.

Evolution of Sustainable Business Strategies

The approach to sustainable business has significantly evolved from the 1990s, when the focus was predominantly on ethical business practices and community investment, which were seen as separate from core business strategies. As environmental awareness increased, sustainability became a strategic business imperative.

The integration of sustainability into core business functions, such as energy management, product innovation, and efficient resource use, has proven crucial for businesses aiming to reduce costs, enhance operational efficiency, and meet regulatory standards. The concept of the "triple bottom line," which considers environmental, social, and economic impacts, has guided companies in broadening their sustainability efforts beyond mere compliance to genuine competitive advantage (Belmonte-Urena et al., 2022).

In recent years, the role of technology in sustainable business practices has become increasingly central. Companies have been leveraging innovations to develop products and services that meet the growing consumer demand for sustainability. This includes significant investments in renewable energy and technologies that minimize environmental degradation (Jin et al., 2022).

Collaboration has also been key, with businesses, governments, and nongovernmental organizations partnering to tackle global sustainability challenges such as climate change and resource scarcity. These partnerships often focus on sharing resources and knowledge to foster innovations that address shared environmental goals.

Education and preparedness for future challenges have become more integrated into business strategies, with companies focusing on equipping future leaders with the knowledge and tools needed to implement sustainable practices effectively.

5.2 *Sustainable Technological Innovation: Shaping Sustainable Business Practices in the Digital Era*

Environmental Footprint Reduction and Circular Economy Transition
Sustainable technological innovation (STI) is poised to significantly reshape business strategies, particularly in the context of challenges such as climate change and resource scarcity. Businesses are increasingly integrating sustainability at the core of their operations, compelled by both environmental concerns and market demands. STI facilitates substantial reductions in environmental footprints through the adoption of cleaner technologies, such as renewable energy sources and smart supply chain management (SCM). Additionally, advanced data analysis and monitoring technologies enable real-time assessments of environmental impacts, empowering more informed decision-making for enhanced sustainability (Long et al., 2023).

A significant advancement driven by STI is the shift toward a circular economy (CE), where businesses strive to minimize waste and maximize resource reuse. Technology plays a pivotal role in identifying recycling and reusing opportunities, fostering circular supply chains, and creating new business avenues through the utilization of recycled materials. Moreover, technologies such as blockchain enhance traceability and transparency in supply chains, improving the verification of ethical and environmental standards. Artificial intelligence (AI) and automation also improve safety and mitigate risks associated with hazardous work environments, thereby enhancing operational efficiency and worker welfare (Shakina et al., 2021).

Enhanced Diversity, Equity, and Inclusion
Technological innovations bolster efforts to enhance diversity and inclusion within workplaces. Advanced data analytics tools help identify and rectify biases in hiring and promotion processes, aiding in the creation of more diverse and inclusive workforces. Technological advancements also promote workplace flexibility, enhancing the inclusion of individuals with disabilities and fostering a more equitable work culture.

Transformed Stakeholder Relationships
Technological innovations significantly influence the dynamics between companies and their stakeholders, including consumers, investors, and communities. Technologies like social media and online platforms facilitate transparent and open communication, fostering trust and enabling direct interaction with consumers. This interaction allows companies to receive real-time feedback, influencing decision-making processes and the continual improvement of products and services.

Enhanced Reporting for Investor Confidence
Sustainability has become a critical factor in investment decisions, and technology plays a crucial role in this realm. Efficient systems for reporting on environmental, social, and governance (ESG) performance aid investors in evaluating and

allocating capital to businesses committed to sustainable practices (Abad-Segura et al., 2019).

Community Engagement and Social Impact
Technology enables businesses to effectively contribute to local communities through support for sustainable development initiatives and collaborative projects. Technologies facilitate the identification of opportunities for employee volunteerism and support for local causes, enhancing community engagement and generating positive social impacts.

5.3 *Research Recommendations*

The dynamic nature of STI necessitates continuous research to adapt to emerging global scenarios and technological advancements. Research should explore novel avenues such as the impact of digital technologies on traditional industries, the integration of sustainable practices within new sectors, and the development of frameworks that support the transition to a digital and sustainable economy. Investigating these areas will help identify trends and emerging themes that are crucial for future technological and business innovations (Waltman & van Eck, 2012).

Technology Gap in SMEs
Understanding the digital divide among SMEs is crucial for their growth and competitiveness. Research should explore whether cultural barriers among managers contribute to technological reluctance or if there should be a shift toward the practical application of technology rather than merely acquiring knowledge. Investigating common misconceptions about technology, such as viewing it solely as a cost-saving tool rather than a means to increase revenue and customer engagement, is essential. Bridging the technology gap in SMEs is crucial for their growth and competitiveness (Vrontis et al., 2021). Technologies like blockchain can enhance traceability and transparency in supply chains, ensuring companies meet ethical and environmental standards. Advanced data monitoring and analysis will also improve efficiency and accountability in supply chain management.

Cryptocurrency and Blockchain Technology
This research area explores the role of cryptocurrencies and how blockchain technology secures transactions, controls asset transfers, and contributes to the Internet of Value (IoV). The study should consider the advantages and disadvantages of cryptocurrencies, including regulatory challenges and speculative use. Additionally, focusing on the impact of this technology on accounting practices and decision-making processes can help in understanding how regulatory landscapes might improve (Choi, 2021; Keuter et al., 2021). Viewing regulation as an opportunity for improvement, the research should insist on how technological innovation (TI) affects accounting practice and the information that facilitates decision-making processes. Interdisciplinary research in the context of globalization responds positively

to the advancement of knowledge, registering these processes in a consensual way to continue providing theoretical and practical tools for business competitiveness (Laux & Stocken, 2018).

Investing in Technology for Social and Environmental Solutions
Investment strategies that emphasize socially responsible investment (SRI) can significantly influence technological innovation. SRI can drive investment into companies seeking technological solutions for social and environmental issues, fostering innovation in areas like renewable energy, digital health, and sustainability. Such investments not only address critical global challenges but also create financial demand for innovation in these sectors, helping companies meet investor expectations and societal needs. For instance, technology companies focused on energy efficiency, carbon emissions reduction, or sustainable resource management can attract investors who are committed to environmental sustainability. This dynamic can enhance research and development in these areas, promoting the adoption of technologies that reduce environmental impact and improve overall sustainability(Weston & Nnadi, 2021).

Strategic Technology Investment and Its Social Impact
This line is constantly evolving as technology and awareness of digital impacts advance. Some emerging areas of research will include the following:

(i) Ethics of AI: Investigate how companies can ensure ethics and transparency in the implementation of AI systems and algorithms, minimizing bias and ensuring fair decisions.
(ii) Data privacy and consumer protection: Study how companies can comply with privacy regulations, ensure consumer data protection and at the same time leverage information responsibly.
(iii) Digital sustainability: Explore how companies can minimize their carbon footprint and reduce the environmental impacts of their digital operations, including data center management and the use of renewable energy.
(iv) Digital inequality: Investigate how CDR strategies can address the digital divide and ensure equitable access to technology and its benefits.
(v) Ethics in data intelligence and surveillance: Analyze the ethical implications of data collection and surveillance, especially in the context of cybersecurity and crime fighting.
(vi) Accountability on social media platforms: Explore how social media and online platforms can more effectively manage hate speech, misinformation, and content moderation.

These technologies contain the potential to revolutionize our lifestyles and professional landscapes, yet concurrently present critical ethical, social, and economic considerations. The escalating influence of AI and automation necessitates a profound understanding from corporations, urging them to confront the challenges inherent in these technologies. Questions such as the dearth of transparency and accountability in AI systems, the potential for bias and discrimination, and the ramifications of AI on employment and the labor market stand out as the key concerns that require thoughtful and proactive treatment (Trittin-Ulbrich & Böckel, 2022).

5.4 *Practical Implications*

The analysis of sustainable technological innovation (STI) seeks to provide useful knowledge to answer, among others, the following questions: How can companies adeptly assess and manage the social, environmental, and economic impacts stemming from their digital technologies and practices? How can we ensure that the benefits of digital technologies are distributed fairly and equitably among different groups and societies? What are the consequences of not addressing corporate digital responsibility (CDR) and how to manage the risks? How to integrate CDR into corporate decision-making processes and communicate effectively to stakeholders? What role can governments, civil society organizations, and other stakeholders play in promoting and implementing CDR?

STI has a number of significant practical implications across a wide spectrum of areas, from environmental management to stakeholder engagement and transparency. STI provides companies with the ability to address critical environmental challenges, resulting in a reduction of their ecological footprint. This includes the adoption of more efficient technologies in the use of resources, waste management, and optimization of the supply chain to minimize environmental impact. Companies can employ sensors and connected devices (IoT) to monitor and optimize energy and water resource consumption in real time, which not only reduces costs but also demonstrates their commitment to sustainability. Additionally, technology enables the transition to renewable energy sources and the implementation of clean energy strategies, supporting carbon emissions reduction goals and promoting a more environmentally responsible corporate image.

In the area of transparency and traceability, technology plays a fundamental role. Blockchain, for example, allows for transparent and secure recording of company transactions and activities, increasing the trust of consumers and stakeholders. This is especially relevant in the supply chain, where distributed ledger technology can be used to track and verify the provenance of products and raw materials, helping to combat irresponsible business practices such as labor exploitation and illegal logging (Chan et al., 2020).

Another important implication of STI is improved communication and engagement with stakeholders. Social media and other digital platforms allow businesses to interact with their customers, employees, investors, and communities more effectively. This creates a two-way feedback channel that encourages participation and allows companies to respond to the concerns and expectations of their stakeholders. Technology also enables the creation of more advanced and visual sustainability reports, which effectively communicate the company's achievements and objectives, which, in turn, strengthens its reputation.

STI is essential for measuring and monitoring the environmental and social impact of companies. Through data collection and analysis, companies can more accurately assess their performance in areas such as waste reduction, energy savings, and community impact. Advanced data analytics tools enable companies to identify trends, identify areas for improvement, and make informed decisions to maximize their contribution to sustainability.

In terms of education and training, technology plays an essential role in promoting digital and ethical responsibility among employees. Companies can use e-learning and virtual reality solutions to provide online training on topics such as cybersecurity, data privacy, and ethics in technology. This improves employee preparedness to face digital challenges and demonstrates the company's commitment to responsible online practices.

STIs also contribute to operational efficiency and cost savings. Automation and AI enable more efficient management of operations and decision-making. This translates into savings in resources, time, and money, which can be reinvested in sustainability initiatives. Additionally, the technology helps companies comply with increasingly strict regulations related to data privacy and cybersecurity, thus avoiding potential penalties and reputational damage.

The CDR also covers cybersecurity. Technology is critical to protecting sensitive customer information and company data. Businesses can use advanced security solutions, such as multi-factor authentication, encryption, and AI-based threat detection, to protect their digital assets. This is essential to prevent data loss, theft of personal information, and cyberattacks that can have a negative impact on customer trust and the company's image.

The ability to adapt and respond to changes in the digital environment is another important implication of STI. As the technological landscape evolves, companies must be prepared to face new digital challenges and risks. Technology provides the flexibility to adjust strategies and policies as needed. The ability to quickly adopt new technology solutions is also essential to keep up with changing consumer expectations and evolving government regulations (Murzaliyeva et al., 2018).

6 Conclusions

The aim of this investigation was to scrutinize the research concerning the nexus between sustainable technological innovation (STI) and business competitiveness. This topic has garnered considerable attention in the research domain, evident from its substantial growth over the last three decades.

STI exerts a substantial influence on business practices, empowering companies to adeptly navigate digital challenges and seize opportunities. Across dimensions encompassing environmental management, sustainability, transparency, stakeholder communication, impact measurement, and cybersecurity, technology emerges as a pivotal catalyst in fostering conscientious and ethical business practices. The ensuing practical implications underscore the imperative of seamlessly integrating STI into business strategies, thereby propelling a positive societal and environmental impact.

Within this framework, technology assumes a pivotal role in steering the transition toward a circular economy (CE), advocating for the recycling and reutilization of products and materials rather than their disposal. Utilizing waste tracking and management systems, propelled by sensors and data analysis, companies can optimize waste management, thus supporting sustainability and waste reduction goals.

Moreover, within the financial domain, technology integration (TI) has influenced the burgeoning growth of technology-centric investment platforms. Companies can use these platforms to discern investment opportunities that align with sustainability criteria, attracting investors who value social and environmental responsibility.

This technological arsenal equips companies with advanced tools to assess risks and opportunities related to their business operations. The synergistic application of data analytics, artificial intelligence (AI), and machine learning (ML) can pinpoint nascent threats and market prospects aligned with sustainability objectives. This adeptness empowers companies to preemptively tailor their strategies and make judicious decisions.

STI furnishes companies with potent instruments to efficaciously tackle environmental, social, and ethical challenges, concurrently germinating business prospects and fortifying stakeholder relationships. Companies assimilating sustainability as an integral facet of their strategy are poised to not only navigate future challenges but also flourish in a milieu where sustainability assumes paramount significance. Concurrently, emergent trends accentuate the escalating prominence of corporate digital responsibility (CDR) in a progressively intertwined and technologically sophisticated world.

Despite the insights garnered, this study acknowledges certain limitations, shaping the outcomes obtained and potentially forming the basis for future publications. These limitations encompass the utilization of the Scopus database, the chosen descriptors, the temporal scope for sample selection, and the statistical methodologies employed for data extraction. In essence, the scientific activity on the intersection of STI and business competitiveness has burgeoned globally in a conducive environment since the advent of the twenty-first century. This confluence of interests underscores the widespread enthusiasm for disseminating research findings, propelling both scientific and business advancements.

References

Abad-Segura, E., & González-Zamar, M.-D. (2020). Research analysis on emerging technologies in corporate accounting. *Mathematics, 8*(9), 1589. https://doi.org/10.3390/math8091589

Abad-Segura, E., Cortés-García, F. J., & Belmonte-Ureña, L. J. (2019). The sustainable approach to corporate social responsibility: A global analysis and future trends. *Sustainability (Switzerland), 11*(19). https://doi.org/10.3390/su11195382

Abad-Segura, E., Morales, M. E., Cortés-García, F. J., & Belmonte-Ureña, L. J. (2020). Industrial processes management for a sustainable society: Global research analysis. *PRO, 8*(5), 631. https://doi.org/10.3390/pr8050631

Abernethy, M. A., & Bouwens, J. (2005). Determinants of accounting innovation implementation. *Abacus, 41*(3), 217–240. https://doi.org/10.1111/j.1467-6281.2005.00180.x

Akter, S., Michael, K., Uddin, M. R., McCarthy, G., & Rahman, M. (2020). Transforming business using digital innovations: The application of AI, blockchain, cloud and data analytics. *Annals of Operations Research*. https://doi.org/10.1007/s10479-020-03620-w

Awawdeh, A. E., Ananzeh, M., El-khateeb, A. I., & Aljumah, A. (2022). Role of green financing and corporate social responsibility (CSR) in technological innovation and corporate environmental performance: A COVID-19 perspective. *China Finance Review International, 12*(2), 297–316. https://doi.org/10.1108/CFRI-03-2021-0048

Babenko, V., Demyanenko, O., Lyba, V., & Feoktystova, O. (2021). Assessment the cost-effectiveness of information support for the business processes of a virtual machine-building enterprise inthe framework of industry 4.0. *International Journal of Engineering, Transactions A: Basics, 34*(1), 171–176. https://doi.org/10.5829/IJE.2021.34.01A.19

Bartolacci, F., Castellano, N. G., & Cerqueti, R. (2015). The impact of innovation on companies' performance: An entropy-based analysis of the STAR market segment of the Italian stock exchange. *Technology Analysis and Strategic Management, 27*(1), 102–123. https://doi.org/10.1080/09537325.2014.952624

Batlles-delafuente, A., Belmonte-ureña, L. J., Plaza-úbeda, J. A., & Abad-segura, E. (2021). Sustainable business model in the product-service system: Analysis of global research and associated eu legislation. *International Journal of Environmental Research and Public Health, 18*(19). https://doi.org/10.3390/ijerph181910123

Belmonte-Urena, L. J., Batlles-Delafuente, A., Abad-Segura, E., & Morales, M. E. (2022). Bioeconomy as a way of development and sustainability: A study focused on the field of water. *IOP Conference Series: Earth and Environmental Science, 987*(1). https://doi.org/10.1088/1755-1315/987/1/012019

Bröring, S., Laibach, N., & Wustmans, M. (2020). Innovation types in the bioeconomy. *Journal of Cleaner Production, 266*, 121939. https://doi.org/10.1016/j.jclepro.2020.121939

Chan, K. C., Fung, H.-G., & Yau, J. (2013). Predominant sources and contributors of influential business ethics research: Evidence and implications from a threshold citation analysis. *Business Ethics: A European Review, 22*(3), 263–276. https://doi.org/10.1111/beer.12024

Chan, H. L., Wei, X., Guo, S., & Leung, W. H. (2020). Corporate social responsibility (CSR) in fashion supply chains: A multi-methodological study. *Transportation Research Part E: Logistics and Transportation Review, 142*, 102063. https://doi.org/10.1016/j.tre.2020.102063

Chenhall, R. H., & Langfield-Smith, K. (1999). The implementation of innovative management accounting systems. *Australian Accounting Review, 9*(19), 37–46. https://doi.org/10.1111/J.1835-2561.1999.TB00116.X

Choi, T. M. T.-M. (2021). Creating all-win by blockchain technology in supply chains: Impacts of agents' risk attitudes towards cryptocurrency. *Journal of the Operational Research Society, 72*(11), 2580–2595. https://doi.org/10.1080/01605682.2020.1800419

Christensen, H. B., & Nikolaev, V. V. (2013). Does fair value accounting for non-financial assets pass the market test? *Review of Accounting Studies, 18*(3), 734–775. https://doi.org/10.1007/s11142-013-9232-0

Comiskey, E. E., & Groves, R. E. V. (1971). The adoption and diffusion of an accounting innovation. *Accounting and Business Research, 2*(5), 67–77. https://doi.org/10.1080/00014788.1971.9728583

Dai, J., & Vasarhelyi, M. A. (2017). Toward blockchain-based accounting and assurance. *Journal of Information Systems, 31*(3), 5–21. https://doi.org/10.2308/isys-51804

de Jesus, A., & Mendonça, S. (2018). Lost in transition? Drivers and barriers in the eco-innovation road to the circular economy. *Ecological Economics, 145*, 75–89. https://doi.org/10.1016/j.ecolecon.2017.08.001

Desplebin, O., Lux, G., & Petit, N. (2021). To be or not to be: Blockchain and the future of accounting and auditing*. *Accounting Perspectives, 20*(4), 743–769. https://doi.org/10.1111/1911-3838.12265

Ding, X., & Yang, Z. (2020). Knowledge mapping of platform research: A visual analysis using VOSviewer and CiteSpace. *Electronic Commerce Research.* https://doi.org/10.1007/s10660-020-09410-7

Englmann, F. C. (1994). A schumpeterian model of endogenous innovation and growth. *Journal of Evolutionary Economics, 4*(3), 227–241. https://doi.org/10.1007/BF01236370

Gardner, E. C., & Bryson, J. R. (2021). The dark side of the industrialisation of accountancy: Innovation, commoditization, colonization and competitiveness. *Industry and Innovation, 28*(1), 42–57. https://doi.org/10.1080/13662716.2020.1738915

Goodman, L., & Palmon, D. (2001). Technological innovations and the work of the accounting historian: Some key issues. *Accounting Historians Journal, 28*(1), 93–110. https://doi.org/10.2308/0148-4184.28.1.93

Gureyev, V. N., & Mazov, N. A. (2022). Bibliometrics as a promising tool for solving publication ethics issues. *Heliyon, 8*(3), e09123. https://doi.org/10.1016/J.HELIYON.2022.E09123

Healy, M., Cleary, P., & Walsh, E. (2018). Innovativeness and accounting practices: An empirical investigation. *Qualitative Research in Accounting and Management, 15*(2), 231–250. https://doi.org/10.1108/QRAM-06-2017-0047

Hoosain, M. S., Paul, B. S., & Ramakrishna, S. (2020). The impact of 4ir digital technologies and circular thinking on the united nations sustainable development goals. *Sustainability (Switzerland), 12*(23), 1–16. https://doi.org/10.3390/su122310143

Hopkins, J. L. J. L. (2021). An investigation into emerging industry 4.0 technologies as drivers of supply chain innovation in Australia. *Computers in Industry, 125*, 103323. https://doi.org/10.1016/j.compind.2020.103323

Hota, P. K., Subramanian, B., & Narayanamurthy, G. (2020). Mapping the intellectual structure of social entrepreneurship research: A citation/co-citation analysis. *Journal of Business Ethics, 166*(1), 89–114. https://doi.org/10.1007/s10551-019-04129-4

Huo, C., Hameed, J., Zhang, M., Bin Mohd Ali, A. F., & Nik Hashim, N. A. A. (2022). Modeling the impact of corporate social responsibility on sustainable purchase intentions: Insights into brand trust and brand loyalty. *Economic Research-Ekonomska Istrazivanja, 35*(1), 4710–4739. https://doi.org/10.1080/1331677X.2021.2016465

Islam, M. A., & Agarwal, N. K. (2020). What is the impact of library and information science on knowledge management research? *VINE Journal of Information and Knowledge Management Systems, 51*(1), 1–26. https://doi.org/10.1108/vjikms-11-2018-0098

Jamali, D., Yianni, M., & Abdallah, H. (2011). Strategic partnerships, social capital and innovation: Accounting for social alliance innovation. *Business Ethics, 20*(4), 375–391. https://doi.org/10.1111/J.1467-8608.2011.01621.X

Jin, C., Shahzad, M., Zafar, A. U., & Suki, N. M. (2022). Socio-economic and environmental drivers of green innovation: Evidence from nonlinear ARDL. *Economic Research-Ekonomska Istrazivanja, 35*(1), 5336–5356. https://doi.org/10.1080/1331677X.2022.2026241

Kar, S. K., Padhi, S. K., & Samantarai, M. (2016). Disruptive innovation and economic value creation: A service sector perspective. *Prabandhan: Indian Journal of Management, 9*(8), 11–25. https://doi.org/10.17010/pijom/2016/v9i8/99773

Ketels, C. H. M. (2006). Michael Porter's competitiveness framework – Recent learnings and new research priorities. *Journal of Industry, Competition and Trade, 6*(2), 115–136. https://doi.org/10.1007/s10842-006-9474-7

Keuter, V., Deck, S., Giesenkamp, H., Gonglach, D., Katayama, V. T., Liesegang, S., Petersen, F., Schwindenhammer, S., Steinmetz, H., & Ulbrich, A. (2021). Significance and vision of nutrient recovery for sustainable city food systems in Germany by 2050. *Sustainability, 13*(19), 10772. https://doi.org/10.3390/su131910772

Klein, D. A. (2009). The strategic management of intellectual capital. *En The Strategic Management of Intellectual Capital.* https://doi.org/10.4324/9780080517926

Laux, V., & Stocken, P. C. (2018). Accounting standards, regulatory enforcement, and innovation. *Journal of Accounting and Economics, 65*(2–3), 221–236. https://doi.org/10.1016/J.JACCECO.2017.11.001

Li, K., Kim, D. J., Lang, K. R., Kauffman, R. J., & Naldi, M. (2020). How should we understand the digital economy in Asia? Critical assessment and research agenda. *Electronic Commerce Research and Applications, 44*, 101004. https://doi.org/10.1016/j.elerap.2020.101004

Li, N., Huang, Q., Ge, X., He, M., Cui, S., Huang, P., Li, S., & Fung, S.-F. (2021). A review of the research progress of social network structure. *Complexity, 2021*, 1–14. https://doi.org/10.1155/2021/6692210

Liu, Z., & Stephens, V. (2019). Exploring innovation ecosystem from the perspective of sustainability: Towards a conceptual framework. *Journal of Open Innovation: Technology, Market, and Complexity, 5*(3), 48. https://doi.org/10.3390/JOITMC5030048

Long, H., Feng, G.-F., & Chang, C.-P. (2023). How does ESG performance promote corporate green innovation? *Economic Change and Restructuring, 56*(4), 2889–2913. https://doi.org/10.1007/s10644-023-09536-2

Mancini, D., Lombardi, R., & Tavana, M. (2021). Four research pathways for understanding the role of smart technologies in accounting. *Meditari Accountancy Research, 29*(5), 1041–1062. https://doi.org/10.1108/MEDAR-03-2021-1258

Marushchak, L., Pavlykivska, O., Liakhovych, G., Vakun, O., & Shveda, N. (2021). Accounting software in modern business. *Advances in Science, Technology and Engineering Systems, 6*(1), 862–870. https://doi.org/10.25046/aj060195

Mishra, S., Tewari, I., & Toosi, S. (2020). Economic complexity and the globalization of services. *Structural Change and Economic Dynamics, 53*, 267–280. https://doi.org/10.1016/j.strueco.2020.03.002

Moher, D., Altman, D. G., Liberati, A., & Tetzlaff, J. (2011). PRISMA Statement. *Epidemiology, 22*(1), 128. https://doi.org/10.1097/EDE.0b013e3181fe7825

Montégu, J. P., Pertuze, J. A., & Calvo, C. (2021). The effects of importing activities on technological and non-technological innovation: Evidence from Chilean firms. *International Journal of Emerging Markets*. https://doi.org/10.1108/IJOEM-05-2020-0534

Morales, M. E., Batlles-delaFuente, A., Cortés-García, F. J., & Belmonte-Ureña, L. J. (2021). Theoretical research on circular economy and sustainability trade-offs and synergies. *Sustainability, 13*(21), 11636. https://doi.org/10.3390/SU132111636

Murzaliyeva, A., Markhayeva, B., Bakhyt, E., & Karshalova, A. (2018). Social return on investment as instrument for assessing socially responsible investing in the Republic of Kazakhstan. *Journal of Applied Economic Sciences, 13*(5), 1357–1365.

Ousama, A. A., Fatima, A. H., & Majdi, A. R. H. (2011). Usefulness of intellectual capital information: Preparers' and users' views. *Journal of Intellectual Capital, 12*(3), 430–445. https://doi.org/10.1108/14691931111154724

Palmié, M., Wincent, J., Parida, V., & Caglar, U. (2020). The evolution of the financial technology ecosystem: An introduction and agenda for future research on disruptive innovations in ecosystems. *Technological Forecasting and Social Change, 151*, 119779. https://doi.org/10.1016/j.techfore.2019.119779

Pavlatos, O., & Kostakis, H. (2018). Management accounting innovations in a time of economic crisis. *Journal of Economic Asymmetries, 18*, e00106. https://doi.org/10.1016/J.JECA.2018.E00106

Petkov, R. (2020). Artificial intelligence (Ai) and the accounting function – A revisit and a new perspective for developing framework. *Journal of Emerging Technologies in Accounting, 17*(1), 99–105. https://doi.org/10.2308/JETA-52648

Phornlaphatrachakorn, K. (2019). Effects of transformational leadership, organisational learning and technological innovation on strategic management accounting in thailand's financial institutions. *Asian Journal of Business and Accounting, 12*(1), 165–188. https://doi.org/10.22452/AJBA.VOL12NO1.6

Plaza-Úbeda, J. A., Abad-Segura, E., de Burgos-Jiménez, J., Boteva-Asenova, A., & Belmonte-Ureña, L. J. (2020). Trends and new challenges in the green supply chain: The reverse logistics. *Sustainability, 13*(1), 331. https://doi.org/10.3390/su13010331

Shakina, E., Parshakov, P., & Alsufiev, A. (2021). Rethinking the corporate digital divide: The complementarity of technologies and the demand for digital skills. *Technological Forecasting and Social Change, 162*, 120405. https://doi.org/10.1016/j.techfore.2020.120405

Sisaye, S., & Birnberg, J. G. (2014). Sociological approaches to organizational learning: Applications to process innovations in management accounting systems. *Advances in Management Accounting, 23*, 1–43. https://doi.org/10.1108/S1474-787120140000023001

Smith, D. J. (2016). The sustainable and green engine (SAGE) – Aircraft engine of the future? *International Journal of Entrepreneurship and Innovation, 17*(4), 256–262. https://doi.org/10.1177/1465750316672601

Stonehouse, G., & Snowdon, B. (2007). Competitive advantage revisited: Michael porter on strategy and competitiveness. *Journal of Management Inquiry, 16*(3), 256–273. https://doi.org/10.1177/1056492607306333

Tikhonov, A., & Zelentsova, L. (2021). Analysis of external and internal factors of business competitiveness. *Quality – Access to Success, 22*(182), 16–19.

Tohănean, D., Buzatu, A. I., Baba, C.-A., & Georgescu, B. (2020). Business model innovation through the use of digital technologies: Managing risks and creating sustainability. *Amfiteatru Economic, 22*(55), 758–774. https://doi.org/10.24818/EA/2020/55/758

Trittin-Ulbrich, H., & Böckel, A. (2022). Institutional entrepreneurship for responsible digital innovation: The case of corporate digital responsibility. *Creativity and Innovation Management, 31*(3), 447–459. https://doi.org/10.1111/caim.12513

Urraca-Ruiz, A., & Laguna-Molina, N. E. (2016). Dynamic technological specialization, aggregated convergence and growth. *International Economics and Economic Policy, 13*(2), 195–221. https://doi.org/10.1007/s10368-014-0297-4

Van Eck, N. J., & Waltman, L. (2010). Software survey: VOSviewer, a computer program for bibliometric mapping. *Scientometrics, 84*, 523–538.

Vrontis, D., Morea, D., Basile, G., Bonacci, I., & Mazzitelli, A. (2021). Consequences of technology and social innovation on traditional business model. *Technological Forecasting and Social Change, 170*, 120877. https://doi.org/10.1016/j.techfore.2021.120877

Waltman, L., & van Eck, N. J. (2012). A new methodology for constructing a publication-level classification system of science. *Journal of the American Society for Information Science and Technology, 63*(12), 2378–2392. https://doi.org/10.1002/asi.22748

Waltman, L., van Eck, N. J., & Noyons, E. C. M. (2010). A unified approach to mapping and clustering of bibliometric networks. *Journal of Informetrics, 4*(4), 629–635. https://doi.org/10.1016/j.joi.2010.07.002

Watari, T., Nansai, K., Nakajima, K., McLellan, B. C., Dominish, E., & Giurco, D. (2019). Integrating circular economy strategies with low-carbon scenarios: Lithium use in electric vehicles. *Environmental Science and Technology, 53*(20), 11657–11665. https://doi.org/10.1021/ACS.EST.9B02872

Weston, P., & Nnadi, M. (2021). Evaluation of strategic and financial variables of corporate sustainability and ESG policies on corporate finance performance. *Journal of Sustainable Finance and Investment, 13*, 1058. https://doi.org/10.1080/20430795.2021.1883984

Williams, M. D. (2005). Measuring the impact of technological innovation on sustainable development in San Diego. *World Review of Science, Technology and Sustainable Development, 2*(1), 11–34. https://doi.org/10.1504/WRSTSD.2005.006725

Witt, U. (2002). How evolutionary is Schumpeter's theory of economic development? *Industry and Innovation, 9*(1–2), 7–22. https://doi.org/10.1080/13662710220123590

Wonglimpiyarat, J. (2016). The new Darwinism of the payment system: Will bitcoin replace our cash-based society? *The Journal of Internet Banking and Commerce, 21*(s2), 1–15. https://doi.org/10.4172/1204-5357.S2-002

Wyatt, A. (2001). Accounting for intangibles: The great divide between obscurity in innovation activities and the balance sheet. *Singapore Economic Review, 46*(1), 83–117. https://doi.org/10.1142/S0217590801000243

Advances Toward Agri-food Sustainability in the European Union: Challenges and Perspectives

Francisco José Castillo-Díaz, Ana Batlles-delaFuente, María J. López-Serrano, and Luis J. Belmonte-Ureña

Contents

1 Introduction

Sustainability has become an increasingly important concern on the international stage, particularly in the agri-food sector due to its integral role in food security and the sovereignty of states (Castillo-Díaz et al., 2023; Orou et al., 2023). With a

F. J. Castillo-Díaz (✉) · A. Batlles-delaFuente · M. J. López-Serrano · L. J. Belmonte-Ureña
Department of Economy and Business, Sustainable Protected Agriculture Research Network, University of Almería La Cañada de San Urbano, Almería, Spain
e-mail: fcastillo@ual.es; anabatlles@ual.es; mjlopezserrano@ual.es; lbelmont@ual.es

S. S. Muthu (ed.), *Assessment of Social Sustainability Management in Various Sectors*, Environmental Footprints and Eco-design of Products and Processes,
https://doi.org/10.1007/978-3-031-81029-9_7

growing recognition of the environmental, social, and economic challenges confronting humanity, it is imperative to address these issues both comprehensively and with urgency. In this regard, the 2030 Agenda, endorsed by the member states of the United Nations (UN), serves as a pivotal framework, outlining 17 Sustainable Development Goals (SDGs) along with 169 specific targets to confront the most pressing global challenges (UN, 2015). Within this framework, the sustainability of activities related to the agri-food sector, including agriculture, livestock, and the agri-food industry, is directly connected to several SDGs, notably the second and 12th. This link between food production and sustainable development underscores the critical need to tackle environmental, social, and economic challenges in the context of food security and long-term sustainability (Bonisoli et al., 2019; Carvalho et al., 2022; Orou et al., 2023; Ruiz-Almeida & Rivera-Ferre, 2019).

The detrimental impacts of human activities on the environment have underscored the necessity of adopting more sustainable practices across all sectors, including the agri-food sector (FAO, 2017a; OCDE-FAO, 2019; UN, 2015). An increased environmental and social consciousness among states has spurred the implementation of policies and actions aimed at mitigating the adverse effects of economic progress and combating climate change. This shift has led countries to revise their structural policies to mitigate their environmental footprint (Cifuentes-Faura, 2022; Davies et al., 2021).

Since the mid-twentieth century, the transformation within agricultural and agri-food industries has precipitated significant environmental impacts globally. The Green Revolution, through genetic enhancements in plant species, not only expanded the Earth's carrying capacity but also intensified production activities, escalating energy and agrochemical demand. Over the past five decades, the consumption of primary production resources has tripled, resulting in more than a 200% increase in cereal productivity since 1961 (FAO, 2017b, 2019a, b). Consequently, this development has led to the emergence of both negative and positive externalities from food production, affecting developed and developing countries alike. Negative impacts include the overexploitation of natural resources, loss of genetic diversity, soil degradation, pollution from zootechnical products, degradation of surface and groundwater quality, ozone layer depletion, and visual pollution of natural landscapes (Gómez-Tenorio et al., 2021; Martínez-Francés et al., 2009; Oehlmann et al., 2009; Pedraza et al., 2015; Zapata Sierra et al., 2022). Furthermore, negative social externalities have been noted, such as land concession rights disputes and the exportation of agri-food products from food-dependent developing countries to economically lucrative markets like Europe. However, positive social impacts, such as enhanced basic services for populations and significant contributions to poverty alleviation, have also been observed. Increased agricultural productivity is correlated with reductions in poverty rates (The World Bank, 2008). The sector's diversification in recent years has facilitated job creation, thereby generating income among the poorest rural populations (Christiaensen et al., 2011; The World Bank, 2008) and enhancing the quality of life in various world regions (OCDE-FAO, 2019).

There is a mounting interest in adopting more sustainable practices within the agri-food sector, signified by a paradigm shift toward reducing the use of agrochemicals and antimicrobials in food production and preserving ecosystem biodiversity (López-Serrano et al., 2020; Pankhurst & Lynch, 1995; Streimikis & Baležentis, 2020; Velten et al., 2015). The European Union has emerged as a key player in initiating environmental protection measures against the impacts of human activities (Cifuentes-Faura, 2022), actively fostering a transition toward a circular and sustainable economy (European Commission, 2020b), as evidenced by the European Green Deal (European Commission, 2019). This strategy aims for climate neutrality by 2050, where net greenhouse gas emissions are zero. To this end, EU member states are mandated to reduce their greenhouse gas emissions by 55% by 2030 (European Parliament, 2021). Additionally, ambitious targets for the agri-food sector have been established, including a 50% reduction in the use of pesticides and zoosanitary products and a 25% reduction in fertilizer use by 2030, alongside a goal to increase the area of crops under European organic certification by 25% (European Commission, 2020a). Biodiversity goals that will affect food production both within and beyond the European Union have also been set (European Commission, 2020c). These environmental commitments present a challenge to the agri-food systems of EU member states, necessitating a change in production methods that have been the standard for over six decades, potentially impacting food sovereignty and security. The availability of synthetic indicators enhances the transition of productive systems by providing strategic information (Carvalho et al., 2022; Castillo-Díaz et al., 2023; Green et al., 2020; Orou et al., 2023).

Given this backdrop, assessing the sustainability of the agri-food sector is of paramount interest. Hence, this chapter aims to explore the tridimensional aspect of sustainability within the European agri-food system, covering agriculture, livestock, and the agri-food industry, to establish a foundation for evidence-based policy decisions.

Sustainability has become an increasingly important concern on the international stage, with the agri-food sector being one of the most relevant areas due to its close relationship with food security and sovereignty of states (Castillo-Díaz et al., 2023; Orou et al., 2023). With growing awareness of the environmental, social, and economic challenges facing humanity, addressing these issues comprehensively and urgently is crucial. In this context, the 2030 Agenda, signed by United Nations (UN) member states, emerges as a crucial framework, establishing 17 Sustainable Development Goals (SDGs) and 169 specific targets to address the most pressing global issues (UN, 2015). Within this agenda, the sustainability of all activities related to the agri-food sector, including agriculture, livestock, and the agri-food industry, is intrinsically linked to various SDGs, such as the 2nnd or 12th SDG. This interconnection between food production and SGDs reflects the critical importance of addressing environmental, social, and economic challenges in the context of food security and long-term sustainability (Bonisoli et al., 2019; Carvalho et al., 2022; Orou et al., 2023; Ruiz-Almeida & Rivera-Ferre, 2019).

The negative effects of human activities on the environment have led to a growing recognition of the need to adopt more sustainable practices in all sectors, including agri-food (FAO, 2017a; OCDE-FAO, 2019; UN, 2015). The emergence of expanded environmental and social awareness among states has driven the adoption of policies and actions aimed at reducing the negative impacts of economic progress and addressing the effects of climate change. Climate change has prompted countries to modify their structural policies to neutralize their environmental footprint (Cifuentes-Faura, 2022; Davies et al., 2021).

The transformation undergone by agricultural and agri-food industries since the mid-twentieth century has generated a series of significant environmental impacts on ecosystems globally. While the Green Revolution expanded the planet's carrying capacity through genetic improvements in plant species, it also intensified production activities and increased the demand for energy and agrochemicals. For example, in the last 50 years, primary production resource consumption has tripled, leading to over a 200% increase in cereal productivity since 1961 (FAO, 2017b, 2019a, b). This scenario has resulted in the emergence of both negative and positive externalities from food production, affecting both developed and developing countries. Among the negative impacts observed are the expansion of natural resource consumption, loss of genetic diversity, soil degradation, soil pollution with zootechnical products, degradation of surface and groundwater quality, ozone layer depletion, and visual pollution of natural spaces (Gómez-Tenorio et al., 2021; Martínez-Francés et al., 2009; Oehlmann et al., 2009; Pedraza et al., 2015; Zapata Sierra et al., 2022). Additionally, negative social externalities such as disputes among residents over land concession rights and the exportation of agri-food products from food-dependent developing countries to economically lucrative markets like Europe have been observed. Positive social impacts can also be observed, such as increased basic services for the population (Ayompe et al., 2021). However, the main benefit lies in poverty alleviation. Increased productivity in agricultural systems is associated with a reduction in the poverty rate (The World Bank, 2008). The dynamics of this sector, which has diversified in recent years, have enabled job creation, promoting income generation among the poorest rural populations (Christiaensen et al., 2011; The World Bank, 2008) and improving the quality of life in many regions of the world (OCDE-FAO, 2019).

In recent years, there has been a growing interest in adopting more sustainable practices in the agri-food sector. This paradigm shift is reflected in the increasing attention paid to reducing agrochemicals and antimicrobials in food production, as well as in preserving and enhancing ecosystem biodiversity (López-Serrano et al., 2020; Pankhurst & Lynch, 1995; Streimikis & Baležentis, 2020; Velten et al., 2015). The European Union, in particular, has been a prominent actor in implementing initiatives to protect the environment from the impacts generated by human activities (Cifuentes-Faura, 2022), actively promoting the transition to a circular and sustainable economy (European Commission, 2020b), as reflected in the European Green Deal (European Commission, 2019). Through this strategy, climate neutrality is sought, meaning that net greenhouse gas emissions are zero by 2050. To achieve this, the European Union has set ambitious environmental and climate goals. By 2030, EU member states are required to reduce their greenhouse gas emissions by 55% (European Parliament, 2021). Additionally, significant targets have been set regarding agri-food. By 2030, a 50% reduction in the application of pesticides and zoosanitary products, as well as a 25% reduction in fertilizer application, is expected. There is also a goal to increase European organic certification crop area by 25% (European Commission, 2020a). Furthermore, biodiversity objectives have been established that will impact food production both within and outside the European Union (European Commission, 2020c). These new environmental agreements pose a challenge to the agri-food systems of EU member states, as they must change the way they produce food, adapting production techniques that have been in place for over six decades and may have negative effects on food sovereignty and security. The availability of synthetic indicators improves the transition of productive systems by providing information for strategy development (Carvalho et al., 2022; Castillo-Díaz et al., 2023; Green et al., 2020; Orou et al., 2023).

In this context, determining the sustainability of the agri-food sector is a matter of special interest. For this reason, the aim of this chapter was to determine the triple dimension of sustainability of the European agrifood system, which encompasses agriculture, livestock and the agrifood industry, to establish a framework for evidence-based policy decisions. In order to meet the stated objective, the book's chapter is divided into four sections after the respective introductory section. Firstly, the methodology used to carry out this work is shown. Secondly, the status of the overall sustainability of the European agri-food sector is presented. Third, the state of the triple bottom line of sustainability (i.e., economic, social, and environmental) of EU food and drink production and processing is presented. Finally, the fourth part presents the main conclusions drawn from this study.

2 Methodology

As previously established, the principal aim of this study was to evaluate the sustainability of the European agri-food sector within the context of the circular economy. This goal was achieved by employing a systematic review process to analyze studies, technical reports, and statistics sourced from both public and private organizations. The methodology adhered to the procedure outlined by Duque-Acevedo et al. (2020). The primary scientific database utilized for this review was Scopus, which is acknowledged as one of the most expansive repositories of scientific literature, encompassing articles, reviews, book chapters, and books (Elsevier, 2023; FECYT, 2022). Furthermore, this search was enriched through the inclusion of literature from public institutions and the consultation of public data repositories. The comprehensive review process was undertaken during the first quarter of 2024, ensuring a thorough and up-to-date analysis (Fig. 1).

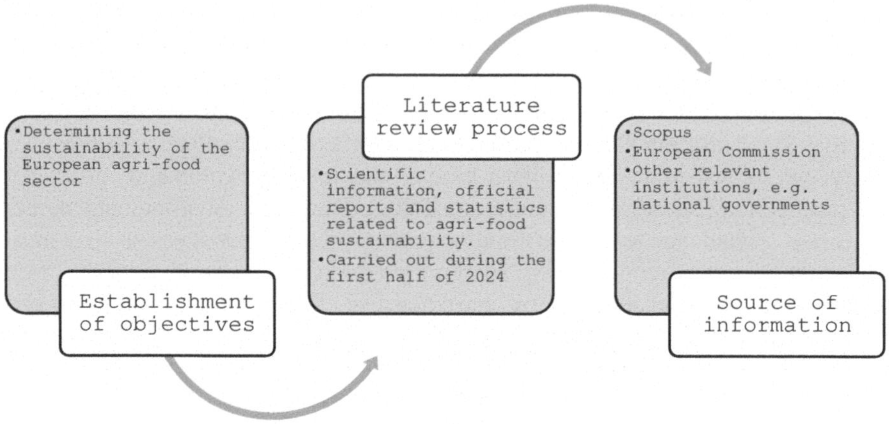

Fig. 1 Phases of the study. (Source: Own elaboration)

3 The Global Sustainability of the European Agri-food Sector

Recent years have seen an increasing focus on the tripartite dimension of sustainability within the European agri-food sector. While the bulk of these studies are theoretical, employing a theoretical basis for the selection of sustainability indicators (Abbate et al., 2023; Correia et al., 2020; Orou et al., 2023; Silvestri et al., 2022a, b), a notable exception is the work of Castillo-Díaz et al. (2023). In their study, a quantitative analysis of sustainability across the 27 EU member states was conducted using a composite indicator that incorporates 49 simple indices. The indicators utilized are detailed in Table 1.

Employing a composite indicator methodology facilitates a holistic evaluation of sustainability, capturing its environmental, social, and economic dimensions. The analysis by Castillo-Díaz et al. (2023) marks a significant contribution by moving beyond theoretical discussions to provide a quantified assessment of the actual sustainability performance of the European agri-food sector.

This study illuminates areas requiring improvement and underscores the variability in sustainability performance across EU member states. By quantitatively measuring sustainability through a multifaceted set of indicators, it enables policymakers to grasp more definitively the challenges and opportunities for bolstering sustainability within the agri-food sector.

In conclusion, empirical research such as that undertaken by Castillo-Díaz et al. (2023) plays a pivotal role in underpinning evidence-based policy formulation and steering the direction of strategies designed to enhance sustainability in the European agri-food sector.

Following the analysis of their results, Castillo-Díaz et al. (2023) reported that the European agri-food sector displays a moderate level of sustainability across economic, social, and environmental dimensions, as illustrated in Fig. 2. France is

Table 1 Indicators used by Castillo-Díaz et al. (2023) to quantify the sustainability of agriculture, livestock, and the agri-food industry in EU member states

	Code	Index	Unit	Impact on sustainability†
Economic subcomponent	E.1	Farm income	Euros per unit of annual work	+
	E.2	Weight of subsidies in farm income	%	−
	E.3	Variability of farm income	%	−
	E.4	Share of agriculture, livestock, and forestry resources in GDP	%	+
	E.5	Weight of the food industry in GDP	%	+
	E.6	Exports	Billions of euros	+
	E.7	Trade balance	Billions of euros	+
	E.8	Agriculture and livestock productivity	Euros per unit of annual work	+
	E.9	Food industry productivity	Euros per person	+
	E.10	Annual consumer price index (CPI)	%	−
	E.11	Protected designation of origin	Number	+
	E.12	Protected geographical indication	Number	+
	E.13	Traditional specialty guaranteed	Number	+

(continued)

Table 1 (continued)

	Code	Index	Unit	Impact on sustainability†
Social subcomponent	S.1	Total, CAP investment	Billions of euros	+
	S.2	Direct payment, CAP	Euros per farm	+
	S.3	Young farmer support, CAP	Euros per farm	+
	S.4	Small producer support, CAP	Euros per farm	+
	S.5	Market measures, CAP	Euros per farm	+
	S.6	Farms subsidized to participate in local market quality programs and short circuits	%	+
	S.7	Farms subsidized to modernize their facilities	%	+
	S.8	Employment in the agricultural and livestock sector	%	+
	S.9	Employment in the forestry sector	%	+
	S.10	Employment in the agri-food industry	%	+
	S.11	Poverty rate in rural areas	%	−
	S.12	Imbalance between the overall poverty rate and the general poverty rate	%	−
	S.13	Generational turnover. Employment in the agricultural, livestock, and forestry sector between 15 and 39 years of age	%	+
	S.14	Generational replacement. Employment in the food industry between 15 and 39 years of age	%	+
	S.15	Producer organizations	Number	+

(continued)

Table 1 (continued)

	Code	Index	Unit	Impact on sustainability[†]
Environmental subcomponent	En.1	Green payment, CAP	Euros per farm	+
	En.2	NH_3 emissions from agriculture and livestock farming	Thousands of tons of NH_3 per farm	–
	En.3	Greenhouse gas emissions per agricultural operation	Tons CO_2 equivalent per agricultural operation	–
	En.4	CO emissions per agri-food industry	Tons of CO per food industry	–
	En.5	Subsidized farms to reduce their NH_3 footprint	%	+
	En.6	Grassland in useful agricultural area	%	+
	En.7	Organic farming	%	+
	En.8	Heads of organic livestock	%	+
	En.9	Organic farming operators	Number	+
	En.10	Water footprint per agricultural holding	Cubic hectometers per agricultural holding	–
	En.11	Agricultural areas subsidized to improve their water efficiency	%	+
	En.12	Agricultural area under crop diversification	%	+
	En.13	Subsidized farms to improve soil quality	%	+
	En.14	Index of birds on farms	%	+
	En.15	Forestry area subsidized to improve soil fertility	%	+
	En.16	Subsidized agricultural land to improve its biodiversity	%	+
	En.17	Sales of antimicrobials in food-producing animals	Milligrams per unit population	–
	En.18	Sale of plant protection products per unit of surface area	Kilograms per hectare	–
	En.19	Consumption of nitrogen fertilizers per unit of surface area	Kg per hectare	–
	En.20	Consumption of phosphate fertilizers per unit of land area	Kg per hectare	–
	En.21	Energy consumption per unit area in agriculture, livestock, and forestry	Euro per kilogram of oil equivalent	–

Source: Own elaboration based on other authors (Castillo-Díaz et al., 2023)

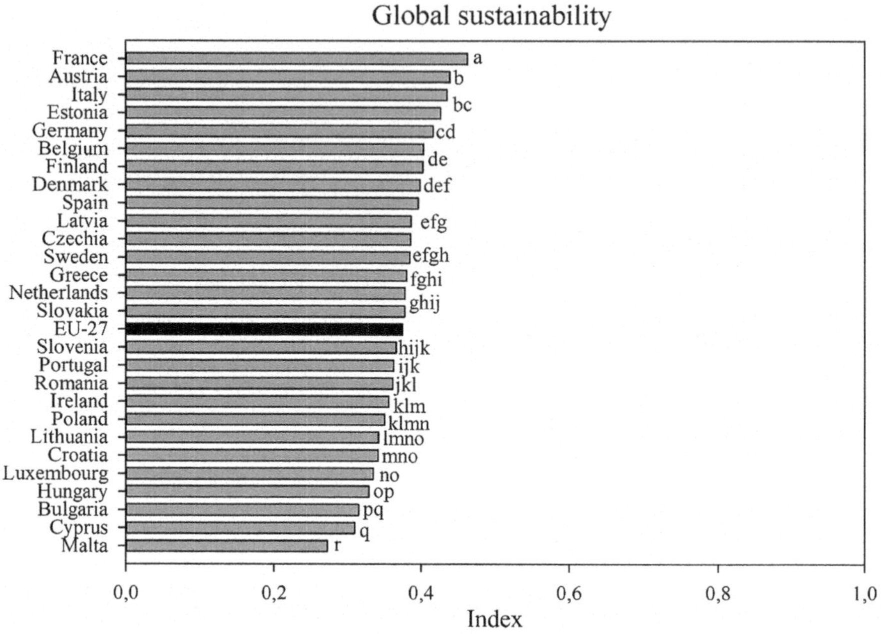

Fig. 2 Global sustainability of the agri-food sector in EU member states (Dimensionless). Note: This indicator was generated from the simple indices shown in Table 1. Before constructing the indicator, the indices have undergone a normalization process to obtain dimensionless units from the maximum and minimum values of the indicators. (Source: Own elaboration based on other authors Castillo-Díaz et al., 2023)

at the forefront, with Austria and Italy closely following. Estonia, Germany, Belgium, Finland, Denmark, Spain, Lithuania, the Czech Republic, Sweden, Greece, the Netherlands, and Slovakia surpass the average sustainability level among EU countries. In contrast, Malta, Cyprus, Bulgaria, and Hungary demonstrated the lowest overall sustainability levels in the agri-food sector. These findings underscore that no EU member state has achieved the theoretical maximum of the model, underlining the imperative for enacting measures to improve sustainability performance within the agri-food sector.

The dimensionless representation of sustainability, as depicted in Fig. 2, provides a detailed comparative analysis of the relative performance of EU member states in the agri-food sector. It highlights the existing disparities among the countries and signals the necessity for targeted interventions to overcome sustainability challenges.

Upon disaggregating the composite indicator, it becomes clear that EU countries achieving significant positions in the ranking depicted in Fig. 2 excel in various dimensions of sustainability. Figure 3 illustrates the outcomes of a cluster analysis that classified EU member states according to their sustainability performance. The determination of the number of clusters was based on the binomial coefficient,

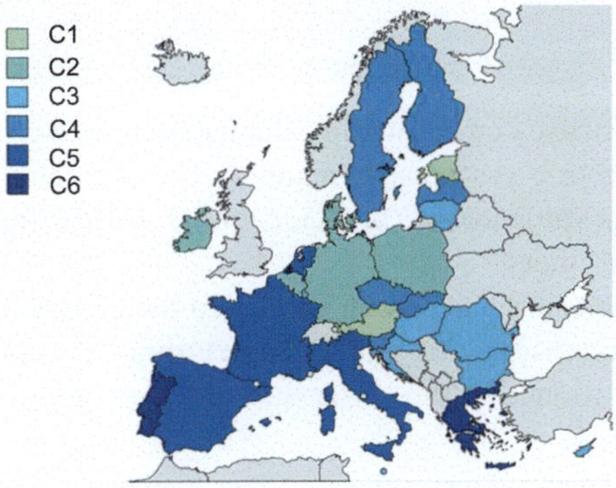

C	SE	SS	SEN
1	0.24	0.26	0.68
2	0.36	0.34	0.44
3	0.26	0.24	0.44
4	0.26	0.29	0.53
5	0.56	0.32	0.40
6	0.34	0.23	0.49

Fig. 3 Classification of agri-food sector sustainability in the 27 EU member states based on the combined status of economic, social, and environmental aspects. Note: For the implementation of this classification, a cluster analysis was performed based on Ward's method. The number of clusters was determined by measuring the binomial coefficient of the three aspects of sustainability. *C* cluster, *SE* economic sustainability, *SS* social sustainability, *SEN* environmental sustainability. (Source: Own elaboration based on other authors Castillo-Díaz et al., 2023)

derived from the three pillars of sustainability. The classification is outlined as follows:

- Group I: Comprises member states with high environmental sustainability.
- Group II: Includes member states that, despite their low average values, exhibit the highest social sustainability.
- Group III: Constituted by member states demonstrating low performance across economic, social, and environmental sustainability.
- Group IV: Consists of member states with moderate levels of social and environmental sustainability.
- Group V: Identifies member states that excel in economic and social sustainability, yet display low environmental sustainability.
- Group VI: Encompasses member states with moderate economic and environmental sustainability, but low social sustainability.

4 State of the Triple Sustainability Dimension

This section presents an isolated examination of the tripartite dimension of sustainability, which includes the economic, social, and environmental facets, alongside selected key indices that are indicative of each domain. The objective of this detailed analysis of each dimension of sustainability is to identify areas that require enhancement and to uncover opportunities for evolving toward more sustainable and resilient agri-food systems

4.1 Economic Dimension

The analysis of economic sustainability focuses on evaluating the capacity of agri-food systems to generate economic value in a sustainable manner. The indicators considered in this evaluation include gross value added (GVA), exports, and the net income of agricultural enterprises, among others.

France, the Netherlands, Spain, and Germany have been identified as leaders in economic sustainability within this dimension. In contrast, Estonia, Latvia, Lithuania, Slovenia, and Slovakia are positioned at the lower end of the economic sustainability rankings, as depicted in Fig. 4.

4.1.1 Gross Value Added and Exports

The leading positions in the ranking presented in Fig. 4 are predominantly occupied by member states that excel in generating gross value added (GVA) from agriculture, livestock farming, the agri-food industry, or distribution sectors. France, Italy, Spain, and Germany emerge as the nations whose agriculture and livestock farming

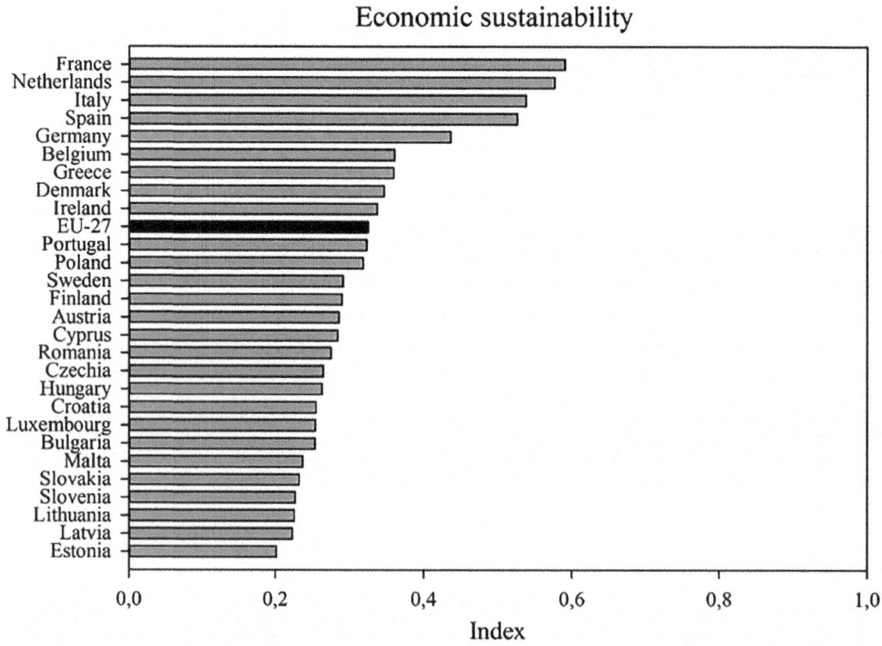

Fig. 4 Economic sustainability of the European agri-food sector (Dimensionless). Note: This indicator was generated from the simple indices shown in Table 1. Before constructing the indicator, the indices have undergone a normalization process to obtain dimensionless units from the maximum and minimum values of the indicators. (Source: Own elaboration based on other authors Castillo-Díaz et al., 2023).

sectors contribute most significantly to their GVA. In the realm of the agri-food industry, Germany stands out as the territory generating the highest GVA, with France, Italy, and Spain following closely behind. In terms of the commercialization sector, the foremost countries are Germany, France, Italy, and Spain, as demonstrated in Fig. 5 (Maudos & Salamanca, 2022, 2023).

4.1.2 Net Entrepreneurial Income from Agricultural Production

Figure 5 illustrates the trend in agricultural profits from 2015 to 2022, revealing that numerous EU agricultural systems have witnessed profits that are similar to or lower than those recorded in 2015. This trend is evident in countries such as Portugal, the Netherlands, Spain, Romania, Slovenia, Slovakia, Hungary, and Malta. Conversely, France has seen its economic profits grow at a rate consistent with inflation. This

Fig. 5 Weight of agri-food production, processing, and marketing of each member state in the sectoral GVA of the EU-27, shown as a percentage. Note: The gross value added (GVA) generated by the agri-food sector is defined as the sum of the GVA generated by agriculture, livestock, and the agri-food industry. The GVA generated by the agri-food sector* is defined as the GVA generated by agriculture, livestock, agri-food industry, and distribution. This classification is based on the indications of other authors (Maudos & Salamanca, 2023)

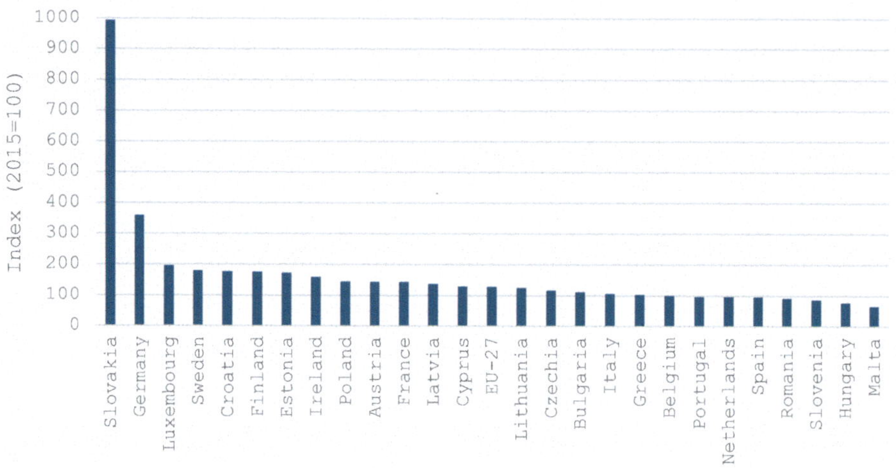

Fig. 6 Net entrepreneurial income from agriculture. (Source: Own elaboration based on EUROSTAT, 2022)

decline in profitability within the primary sector poses a threat to the economic dimension of sustainability, attributable to rising production costs coupled with the reduction or stabilization of source prices (Castillo-Díaz et al., 2022; Honoré et al., 2019). Similarly, the agriculture and livestock industry has faced an increase in production costs, consequently diminishing its profit margins (Fernández et al., 2023) (Fig. 6).

4.2 *Social Dimension*

Social sustainability in the agri-food sector encompasses aspects related to human well-being, equity, inclusion, and social cohesion. The evaluation of this dimension incorporates key indicators such as employment in the agricultural sector, labor conditions, access to nutrition, and the quality of life in rural communities.

The social significance of the agri-food sector is its contribution to maintaining territorial balance across the regions of EU member states, especially in rural areas. Activities such as agriculture, livestock farming, the agri-food industry, and the related auxiliary industries are pivotal in generating employment and fostering economic prosperity in rural locales, which are often afflicted by depopulation issues (Egea et al., 2017; Gómez-Tenorio et al., 2021). Figure 7 presents the social sustainability rankings of the agri-food sector across the 27 EU member states. France, Belgium, Germany, the Czech Republic, Denmark, Finland, Poland, and Spain occupy the top positions in the ranking, whereas Malta, Portugal, Cyprus, and Slovenia are found at the lower end.

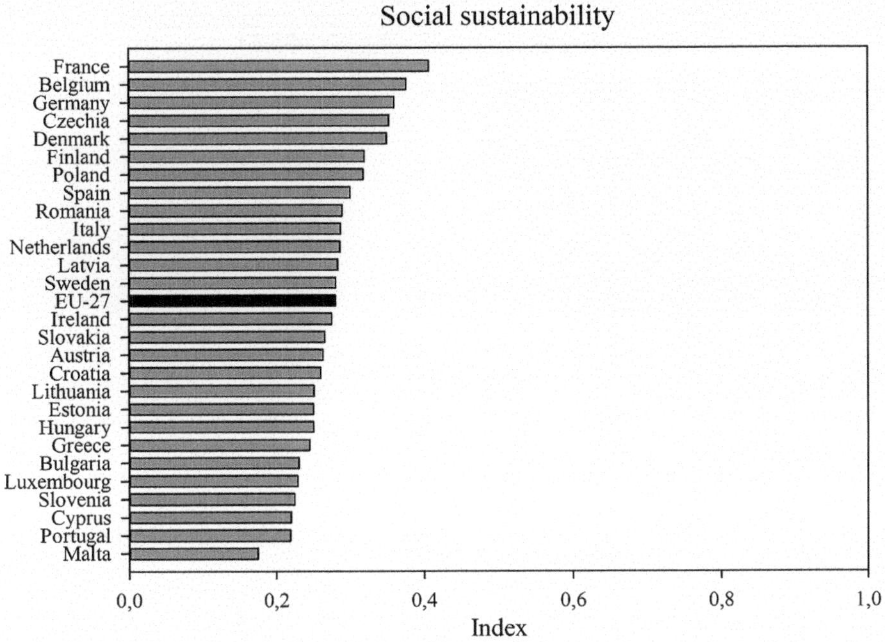

Fig. 7 Social sustainability of the European agri-food sector (Dimensionless). Note: This indicator was calculated based on the simple indices shown in Table 1. Before constructing the indicator, the indices have undergone a normalization process to obtain dimensionless units from the maximum and minimum values of the indicators. (Source: own elaboration based on other authors Castillo-Díaz et al., 2023)

4.2.1 Employment Generated by the Agri-food Sector

A critical aspect within the social dimension of the agri-food sector is the employment it generates. In the European Union, the sector is responsible for creating 46.9 million direct jobs across production, processing, and marketing of food. Notably, primary production constitutes 40.4% of employment within the extended agri-food sector, while marketing accounts for 38.9% of the total employment. Romania stands out as the member state most dependent on the agri-food sector for employment, with nearly 33% of its workforce engaged in this sector. Following Romania, Bulgaria, Greece, Latvia, Poland, Croatia, Portugal, Lithuania, Spain, Cyprus, and Slovenia have significant portions of their employment linked to agri-food activities. In contrast, Luxembourg exhibits a unique stance in this context (Maudos & Salamanca, 2022, 2023) (Fig. 8).

Below is the employment generated by agriculture and the agri-food industry, as well as by the primary production in rural areas.

The analysis reveals a general decline in employment within sectors associated with agriculture and primary production across many EU member states during the period under study. This trend may be attributed to a variety of factors, including the

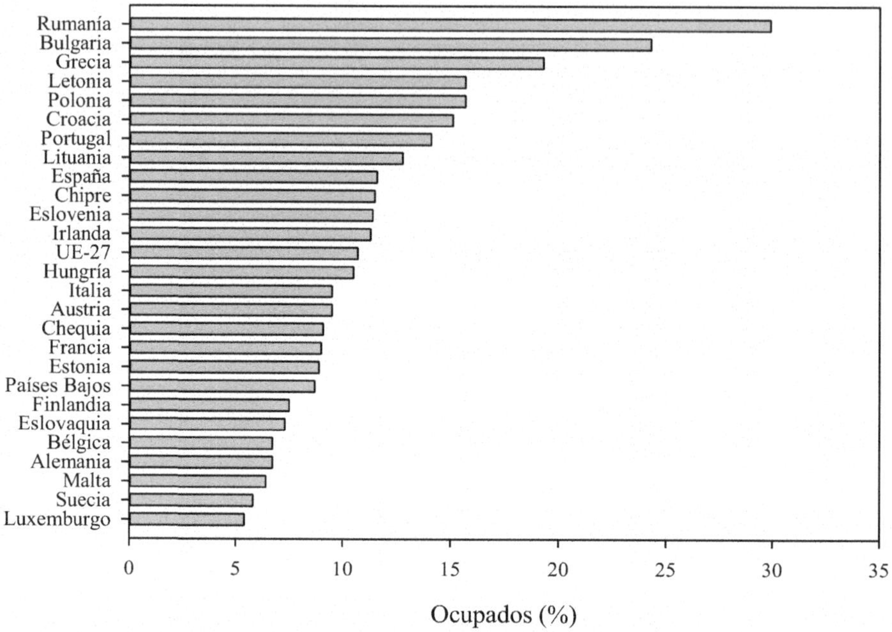

Fig. 8 Illustrates the proportion of employees in the agri-food sector relative to the total employment across EU member states, expressed as a percentage. (Source: Own elaboration based on other authors Joaquín Maudos & Salamanca, 2023)

diversification of economic activities within these countries and the increasing mechanization and automation of primary production processes (Dixon et al., 2001; MAPA & Cajamar, 2022, 2023; Mundial, 2021).

Despite this overarching trend of decline, certain countries exhibit a continued high reliance on agriculture and primary production as significant sources of employment. Notably, Bulgaria, Romania, and Greece persist in maintaining substantial employment levels within these sectors, indicating a distinct economic structure in comparison to other EU nations. As these economies evolve, there is an observable shift toward greater specialization in nonagricultural sectors such as manufacturing, technology, and services, suggesting a transitionary phase in the economic development of these countries (Dixon et al., 2001; Mundial, 2021) (Tables 2, 3, 4, 5, and 6).

4.2.2 Imbalance Between Rural and Urban Poverty Rates

The variability in total poverty rates across EU member states is noteworthy. Austria is distinguished by its relatively low poverty rates, which have demonstrated a consistent decline over the years. In stark contrast, Bulgaria and Romania present significantly higher poverty rates, albeit with a slight decreasing trend observed in

Table 2 Employment generated by agriculture and livestock farming in each member state of the EU, expressed as a percentage

	2015	2016	2017	2018	2019	2020	2021
Austria	4.30	4.10	3.69	3.44	3.33	3.53	3.44
Belgium	1.13	1.21	1.11	0.95	0.90	0.89	0.91
Bulgaria	5.85	5.69	5.92	5.67	5.67	5.65	5.42
Croatia	8.06	6.33	5.67	5.31	5.25	5.45	5.73
Cyprus	3.60	3.33	2.29	1.85	2.06	2.42	2.46
Czechia	2.27	2.30	2.17	2.15	2.10	1.94	1.94
Denmark	2.29	2.13	1.91	1.94	1.97	1.91	1.82
Estonia	2.70	2.79	2.68	2.42	2.08	2.07	1.65
European Union	4.69	4.43	4.30	4.15	4.00	3.96	3.46
Finland	3.09	2.91	2.86	2.82	2.71	2.63	3.03
France	2.49	2.61	2.38	2.29	2.31	2.13	2.25
Germany	1.29	1.21	1.18	1.15	1.11	1.09	1.10
Greece	12.37	11.87	11.67	11.83	11.10	10.19	10.79
Hungary	4.25	4.44	4.46	4.27	4.25	4.29	3.96
Ireland	5.04	5.00	4.73	4.52	4.21	4.21	4.22
Italy	3.44	3.56	3.46	3.44	3.57	3.69	3.77
Latvia	5.56	5.75	5.11	4.86	4.98	5.16	4.85
Lithuania	7.89	6.90	6.71	6.23	5.50	5.01	4.63
Luxembourg	0.93	0.92	1.14	0.93	0.62	0.72	1.01
Malta	1.11	0.82	0.68	0.71	0.75	0.77	0.52
Netherlands	2.09	2.02	1.99	1.84	1.82	1.81	2.13
Poland	11.00	10.01	9.59	9.07	8.68	9.04	8.08
Portugal	6.96	6.37	5.84	5.50	5.03	4.93	2.29
Romania	24.95	22.51	22.2	21.68	20.53	19.86	11.07
Slovakia	2.39	2.11	1.92	1.71	2.06	1.92	1.77
Slovenia	6.55	4.62	5.15	5.13	3.90	3.84	3.91
Spain	3.80	3.88	3.98	3.85	3.69	3.63	3.74
Sweden	1.39	1.27	1.21	1.10	1.12	1.15	1.17

Source: Own elaboration based on EUROSTAT (2024)

certain years. Additionally, countries such as Greece, Italy, and Spain are characterized by considerable poverty rates, experiencing only minor fluctuations over time.

When examining the poverty rates specifically within rural areas, similar patterns are observed across some countries, though these rates are generally elevated in comparison to the aggregate poverty rate. Notably, in countries like Bulgaria, Romania, and Greece, rural poverty rates surpass 40%, underscoring a profoundly challenging situation in these regions. On a more positive note, countries such as Estonia, Finland, and the Netherlands report lower poverty rates in rural areas, indicative of improved access to resources and opportunities in these locales.

Table 3 Employment generated by the agri-food industry in the member states of the EU, expressed as a percentage

	2015	2016	2017	2018	2019	2020	2021	2022
Austria	1.71	1.96	1.96	1.87	1.89	1.85	1.98	2.17
Belgium	2.40	2.61	2.49	2.42	2.47	2.44	2.23	2.37
Bulgaria	3.64	3.61	3.70	3.50	3.32	3.21	3.13	3.38
Croatia	3.89	3.95	3.50	3.92	4.05	3.53	3.41	3.49
Cyprus	2.99	2.62	2.48	2.10	2.47	2.59	2.39	1.98
Czechia	2.39	2.54	2.34	2.33	2.42	2.31	2.16	2.37
Denmark	2.38	2.11	1.98	2.03	2.01	2.10	2.03	2.04
Estonia	2.54	2.24	2.43	2.27	2.31	2.35	2.22	2.23
European Union	2.44	2.44	2.42	2.40	2.38	2.29	2.27	2.28
Finland	1.57	1.50	1.42	1.51	1.47	1.40	1.52	1.59
France	2.48	2.44	2.47	2.47	2.50	2.37	2.27	2.31
Germany	2.26	2.21	2.18	2.16	2.07	1.71	1.80	1.78
Greece	3.40	3.50	3.65	3.48	3.43	3.22	3.62	3.85
Hungary	3.23	3.25	3.27	3.21	3.18	3.15	3.05	2.84
Ireland	2.73	2.78	2.63	2.55	2.40	2.39	2.41	2.27
Italy	2.15	2.17	2.16	2.23	2.18	2.20	2.19	2.15
Latvia	2.93	3.12	2.69	2.66	2.21	2.28	2.40	2.36
Lithuania	3.19	3.13	3.02	3.07	2.93	2.84	2.89	2.86
Luxembourg	0.35	0.50	0.37	0.61	0.93	0.55	0.26	0.35
Malta	2.57	2.21	1.90	1.97	2.08	1.84	1.74	1.69
Netherlands	1.63	1.67	1.72	1.62	1.47	1.49	1.35	1.39
Poland	3.36	3.35	3.26	3.16	3.20	3.24	3.21	3.13
Portugal	2.30	2.25	2.17	2.22	2.41	2.34	2.16	2.47
Romania	2.30	2.56	2.69	2.67	2.56	2.59	2.92	2.88
Slovakia	2.20	2.18	2.00	1.98	2.19	2.13	1.85	2.12
Slovenia	1.97	1.86	2.15	1.99	2.27	2.09	1.82	1.83
Spain	2.65	2.63	2.69	2.61	2.64	2.72	2.57	2.62
Sweden	0.98	0.95	0.89	0.89	0.88	0.83	0.79	0.83

Source: Own elaboration based on EUROSTAT (2024)

4.3 *Environmental Dimension*

Environmental sustainability critically examines the environmental impact of agri-food activities and assesses the sector's capability to conserve natural resources, reduce pollution, and promote sustainable agricultural practices. Key indicators for this evaluation include carbon footprint, water resource consumption, biodiversity preservation, and effective waste management.

The imperative to measure the environmental dimension arises from an increasing recognition of the need to integrate economic activities with sustainable development principles. It is crucial to note that transitioning to sustainable production methods may present challenges for many businesses, including those in the

Table 4 Employment generated by primary production in rural areas of the EU, expressed as a percentage

	2016	2017	2018	2019	2020
Austria	8.2	7.8	7.2	6.8	7.1
Belgium	3.5	3.4	3.3	3.2	3.3
Bulgaria	33.3	33.8	32.2	30.7	31.1
Croatia	13.8	12.7	11.4	11.1	10.9
Czechia	5.5	5.4	5.4	5.2	5.4
Denmark	4.8	4.7	4.7	4.6	4.4
Estonia	8.2	7.7	7.5	7.3	6.5
European Union	12.7	12.4	12.1	NA	11.6
Finland	6.5	6.3	6.0	5.9	6.1
France	5.6	5.5	5.5	5.5	5.5
Germany	3.2	3.2	3.1	3.0	2.9
Greece	23.6	23.8	23	22.3	22.1
Hungary	7.3	7.4	7.3	7.2	7.3
Ireland	8.7	8.4	7.9	7.4	7.6
Italy	7.9	7.8	8.0	7.9	7.9
Latvia	16.5	16.6	16.3	16.8	18.2
Lithuania	25.9	27.2	23.8	18.6	18.6
Netherlands	4.5	4.5	4.0	3.9	3.9
Poland	21.0	20.3	19.0	18.3	18.9
Portugal	20.2	19.4	18.7	17.4	17.4
Romania	33.5	33.1	33.2	32.2	31.0
Slovakia	4.1	4.0	3.9	3.8	3.8
Slovenia	11.4	11.0	10.6	10.4	10.2
Spain	11.9	11.5	11.1	10.8	10.7
Sweden	5.0	4.7	4.6	4.6	4.5

Source: Own elaboration based on EUROSTAT (2024). *NA* not available

agri-food sector. These challenges could lead to difficulties in adjusting to the evolving market conditions, potentially resulting in operational cessation, economic loss, and job redundancies. Moreover, within sectors such as agri-food, these challenges may jeopardize food security and sovereignty of nations striving for sustainability (Castillo-Díaz et al., 2023; Cifuentes-Faura, 2022; Kirchherr et al., 2017).

Figure 9 presents the environmental sustainability ranking of the agri-food sector across EU member states. Estonia, Austria, and Latvia are distinguished by their leading positions in environmental sustainability, whereas the Netherlands, Malta, and Spain are identified as having the least sustainable practices within the sector.

Below are the key environmental indices that play a crucial role in assessing the sustainability of the European agri-food sector:

Table 5 Total poverty rate in the territories of the EU, expressed as a percentage

	2016	2017	2018	2019	2020
Austria	8.2	7.8	7.2	6.8	7.1
Belgium	3.5	3.4	3.3	3.2	3.3
Bulgaria	33.3	33.8	32.2	30.7	31.1
Croatia	13.8	12.7	11.4	11.1	10.9
Czechia	5.5	5.4	5.4	5.2	5.4
Denmark	4.8	4.7	4.7	4.6	4.4
Estonia	8.2	7.7	7.5	7.3	6.5
European Union	12.7	12.4	12.1	NA	11.6
Finland	6.5	6.3	6.0	5.9	6.1
France	5.6	5.5	5.5	5.5	5.5
Germany	3.2	3.2	3.1	3.0	2.9
Greece	23.6	23.8	23	22.3	22.1
Hungary	7.3	7.4	7.3	7.2	7.3
Ireland	8.7	8.4	7.9	7.4	7.6
Italy	7.9	7.8	8.0	7.9	7.9
Latvia	16.5	16.6	16.3	16.8	18.2
Lithuania	25.9	27.2	23.8	18.6	18.6
Netherlands	4.5	4.5	4.0	3.9	3.9
Poland	21.0	20.3	19.0	18.3	18.9
Portugal	20.2	19.4	18.7	17.4	17.4
Romania	33.5	33.1	33.2	32.2	31.0
Slovakia	4.1	4.0	3.9	3.8	3.8
Slovenia	11.4	11.0	10.6	10.4	10.2
Spain	11.9	11.5	11.1	10.8	10.7
Sweden	5.0	4.7	4.6	4.6	4.5

Source: Own elaboration based on EUROSTAT (2024). *NA* not available

4.3.1 Greenhouse Gas Emissions

Table 7 presents an in-depth look at the greenhouse gas (GHG) emissions within the agricultural sector across EU member states from 2011 to 2021.

The data reveals considerable variability in the share of GHG emissions attributed to agriculture among these countries. Notably, Ireland and Denmark have persistently reported high percentages, exceeding 20%, in contrast to Malta and Luxembourg, where agricultural emissions constitute less than 5% of their total emissions.

A critical observation is the temporal trend of these emissions. Over the years, a gradual uptick in the proportion of emissions from agriculture has been recorded in several countries, with Spain, Portugal, and Romania experiencing notable increases. This trend indicates the challenges these nations face in shifting toward more sustainable agricultural practices. Conversely, Estonia and Latvia have experienced significant annual fluctuations, potentially due to climatic variations, changes in

Table 6 Poverty rate in rural areas of EU member states expressed as a percentage

	2015	2016	2017	2018	2019	2020	2021
Austria	13.9	14.8	13.0	14.2	11.3	11.3	12.6
Belgium	19.1	19.6	21.0	18.8	18.6	17.3	17.5
Bulgaria	54.8	53.8	51.9	47.4	48.5	46.7	40.3
Croatia	34.9	33.5	33.1	30.9	28.7	28.0	NA
Cyprus	33.1	32.2	29.5	26.4	24.9	28.7	30.8
Czechia	12.8	11.4	11.6	11.6	11.9	11.7	NA
Denmark	15.8	13.4	13.1	15.1	14.3	12.7	NA
Estonia	26.4	26.1	26.9	29.5	28.5	27.1	22.8
European Union	25.9	26.1	24.4	23.8	22.4	22.8	22.2
Finland	17.2	17.1	15.6	16.6	16.3	17.5	15.5
France	15.9	15.8	13.7	13.7	13.9	13.1	14.4
Germany	16.9	17.2	16.8	17.5	15.6	18.9	17.4
Greece	38.6	38.9	38.9	35.2	34.4	32.4	35.2
Hungary	32.5	31.3	31.0	25.8	23.3	22.2	24.1
Ireland	26.5	24.4	22.8	21.5	21.1	20.1	24.5
Italy	26.8	31.0	28.7	26.7	25.2	27.8	25.2
Latvia	37.0	35.0	33.7	32.3	32.3	29.4	NA
Lithuania	34.7	37.6	37.2	35.4	31.2	29.9	29.0
Luxembourg	15.7	14.3	17.4	17.1	18.5	20.0	16.5
Malta	35.5	2.9	NA	NA	NA	NA	17.9
Netherlands	12.8	12.4	13	12.8	10.9	11.8	13.5
Poland	30.0	27.9	24.2	25.3	24.1	24.2	23.8
Portugal	30.0	29.0	27.5	26.3	25.5	24.4	NA
Romania	50.8	51.7	48.5	45.5	44.3	45.4	45.2
Slovakia	20.6	21.8	18.3	19.1	18.4	17.2	NA
Slovenia	19.4	NA	17.7	16.6	14.2	14.5	13.2
Spain	34.0	33.5	31.1	31.0	28.8	29.1	26.9
Sweden	18.4	18.0	20.6	20.4	20.6	19.3	17.4

Source: Own elaboration based on EUROSTAT (2022). *NA* not available

agricultural policies, or shifts in food production and consumption patterns. It is essential to clarify that these fluctuations do not necessarily equate to an outright increase in GHG emissions, which, on the whole, have seen a decline in recent years.

The reduction of agricultural emissions is paramount for achieving the EU Climate Law's objective of cutting GHG emissions by 55% relative to the 1990 levels (European Parliament, 2021). The agricultural sector's role as a carbon sink, through carbon sequestration, is also noteworthy. Initiatives such as the promotion of low-carbon agriculture and organic farming are gaining traction for their lower emission profiles compared to conventional methods (European Commission, 2020a). Remarkably, 55% of GHG emissions in certain agricultural practices occur before food production begins, during the manufacturing of inputs (Paris et al., 2022). Moreover, agricultural techniques that enhance crops' carbon sequestration capacity, like the use of cover crops, are being increasingly advocated (Gómez-Tenorio et al., 2021).

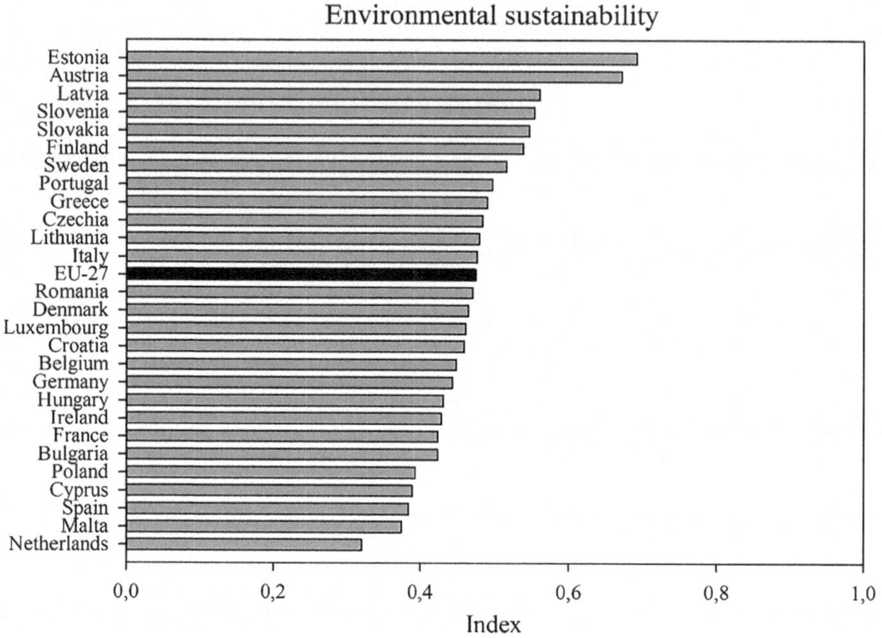

Fig. 9 Environmental sustainability of the European agri-food sector. (Source: Own elaboration based on Castillo-Díaz et al., 2023)

4.3.2 Fertilizer Consumption

Tables 8 and 9 present the consumption of nitrogenous and phosphatic fertilizers across EU member states from 2013 to 2021, underscoring the significant role these nutrients play as primary pollutants in ecosystems.

Germany emerges as a leading consumer of nitrogenous fertilizers, a trend likely reflective of its extensive agricultural practices, robust economy, and pivotal role in European agriculture (Maudos & Salamanca, 2023). Spain, France, and Poland also report substantial nitrogenous fertilizer consumption, indicative of their sizable agricultural sectors. Conversely, smaller nations like Malta and Luxembourg exhibit markedly lower consumption levels.

Temporal analysis unveils a general downward trend in nitrogenous fertilizer consumption, potentially attributable to the adoption of more efficient and sustainable agricultural practices and the enforcement of stricter environmental regulations. This trend is further supported by the growth of organic agriculture, which eschews synthetic fertilizers. However, this decline is not consistent across all member states, highlighting the necessity for tailored national policies to tackle environmental and sustainability challenges (Castillo-Díaz et al., 2023; European Commission, 2020a).

The patterns of phosphatic fertilizer consumption also offer insight. Germany leads in this domain as well, with Spain and France following. Given the critical

Table 7 Proportion of GHG emissions attributable to the primary sector within the total emissions of each EU member state, expressed as a percentage

	2011	2012	2013	2014	2015	2016	2017	2018	2019	2020	2021
EU-27	9.0	9.2	9.4	9.9	9.8	9.8	9.8	9.9	10.2	11.4	10.7
Belgium	7.6	7.7	7.7	8.2	7.9	7.9	7.9	7.8	7.9	8.6	8.1
Bulgaria	7.5	8.3	9.9	10.1	9.7	10.6	10.2	10.7	11.0	12.3	11.2
Czechia	5.4	5.6	5.7	5.9	6.3	6.4	6.2	6.1	6.3	6.8	6.6
Denmark	19.5	21	20.4	22.1	23	22.4	23.5	23.2	24.9	28.3	26.8
Germany	6.4	6.4	6.3	6.8	6.7	6.7	6.7	6.8	7.1	7.7	7.2
Estonia	6.3	7.0	6.6	7.1	8.1	7.2	7.0	7.3	10.6	13.7	12.4
Ireland	30.9	31.9	33.2	32.6	32.1	31.8	32.9	33.9	33.8	36.7	36.2
Greece	7.5	7.7	8.3	8.1	8.3	8.6	8.2	8.4	9.0	10.6	10.1
Spain	8.9	8.8	9.4	9.7	9.5	9.9	9.7	9.8	10.3	12.4	11.6
France	14.4	14.3	14.2	15.5	15.3	15.1	15.0	15.4	15.3	16.8	15.6
Croatia	11.3	11.6	11.1	11.2	11.2	10.9	10.9	10.9	10.7	11.2	10.9
Italy	6.3	6.6	7.0	7.2	7.0	7.3	7.3	7.2	7.4	8.6	7.7
Cyprus	5.7	5.8	5.9	5.5	5.5	5.4	5.4	5.5	5.7	6.8	6.7
Latvia	16.5	17.4	18.1	19.0	19.4	19.4	19.4	17.8	18.9	21.1	20.5
Lithuania	19.8	20.1	21.1	22.2	22.2	21.6	21.3	20.9	20.8	22.1	21.1
Luxembourg	4.9	4.9	5.2	5.5	5.8	6.0	5.8	5.6	5.5	6.6	6.2
Hungary	8.8	9.4	10.5	10.8	10.6	11.0	10.6	10.6	10.6	11.3	11.1
Malta	2.5	2.5	2.7	2.7	3.4	3.8	3.4	3.3	3.2	3.9	3.7
Netherlands	8.5	8.6	8.8	9.3	9.2	9.3	9.4	9.4	9.5	10.7	10.3
Austria	8.6	8.8	8.8	9.3	9.1	9.1	8.8	9.0	8.7	9.6	9.2
Poland	7.8	8.0	8.2	8.4	8.2	8.1	8.1	8.2	8.4	9.1	8.5
Portugal	9.5	9.8	10.1	10.3	9.7	10.0	9.4	9.9	10.6	12.2	12.4
Romania	13.6	13.7	15.5	15.7	16.1	16.2	15.9	16.3	16.6	16.9	16.6
Slovenia	8.8	9.0	9.3	10.5	10.6	10.2	10.0	10.0	10.5	11.3	11.2
Slovakia	5.6	5.7	6.1	6.7	6.1	6.3	5.8	5.9	6.3	6.8	5.9
Finland	9.0	9.7	9.8	10.6	11.2	10.8	11.1	10.8	11.6	13.2	12.9
Sweden	10.8	11.2	11.6	12.1	12.1	12.0	12.4	12.2	12.6	14.4	13.7

Source: Own elaboration based on EUROSTAT (2022)

role of phosphatic fertilizers in plant development, flowering, and fruit formation, their use is essential for agricultural productivity (Camacho-Ferre, 2004). The consumption variability observed among different years and countries may result from shifts in agricultural practices, demand for specific crops, and price fluctuations of fertilizers (Castillo-Díaz et al., 2023).

4.3.3 Pesticide Use

Table 10 delineates the total pesticide consumption across EU member states, measured in kilograms per annum. This table underscores the pronounced disparities in pesticide use among countries, alongside temporal fluctuations. Notably, Germany

Table 8 Consumption of nitrogenous fertilizers per EU member state (tons)

	2013	2014	2015	2016	2017	2018	2019	2020	2021
Belgium	149,743	153,299	150,320	160,883	155,829	149,878	153,353	152,371	152,461
Bulgaria	258,856	322,004	341,608	365,913	351,120	339,329	352,486	364,335	342,890
Czechia	331,616	325,667	397,136	407,184	397,586	351,780	332,032	285,436	309,646
Denmark	193,600	186,800	205,300	240,726	236,492	224,387	225,240	233,475	202,178
Germany	1,648,828	1,675,289	1,822,791	1,710,616	1,658,837	1,496,649	1,342,284	1,372,084	1,265,477
Estonia	33,659	35,806	36,276	36,390	37,333	38,867	41,438	41,486	46,767
Ireland	353,044	331,782	330,959	339,104	369,089	408,495	367,364	379,519	399,164
Greece	182,534	165,942	164,334	185,019	192,167	179,436	189,748	202,945	208,752
Spain	961,507	1,101,895	1,068,103	982,155	1,072,125	1,033,494	1,011,251	1,059,299	1,029,913
France	2,084,375	2,167,376	2,199,550	2,214,700	2,234,189	2,250,835	2,125,147	2,016,775	1,958,478
Croatia	77,920	73,680	87,428	72,320	98,412	99,420	97,099	98,964	102,190
Italy	582,362	581,063	579,728	578,438	577,234	575,857	574,530	573,473	572,105
Cyprus	7051	6693	7533	8073	7841	7824	7824	NA	NA
Latvia	69,696	72,935	75,820	78,285	77,441	74,540	80,718	84,346	84,646
Lithuania	155,000	162,000	166,611	160,237	167,149	159,433	178,603	185,779	187,506
Luxembourg	13,366	12,714	13,007	13,731	13,575	13,038	13,795	13,092	13,932
Hungary	335,515	341,199	378,337	403,970	424,476	424,277	415,901	445,152	456,306
Malta	571	594	618	568	585	586	586	NA	NA
Netherlands	216,026	213,200	244,893	229,825	230,232	212,318	214,752	217,427	211,848
Austria	110,626	121,562	124,078	132,031	111,884	100,096	102,812	117,321	94,319
Poland	1,179,147	1,098,455	1,003,597	1,043,004	1,150,601	1,178,764	994,134	1,033,046	NA
Portugal	111,077	131,643	117,906	108,440	102,584	101,365	107,394	103,171	94,864
Romania	344,468	303,562	357,352	344,311	381,342	468,639	455,964	468,891	510,802
Slovenia	27,263	28,612	28,319	27,095	27,084	27,293	28,048	27,692	29,143
Slovakia	113,581	119,036	114,773	126,236	122,541	128,977	128,533	127,677	127,495
Finland	138,136	147,373	143,479	138,128	138,948	138,385	146,798	139,316	145,807
Sweden	161,100	181,100	190,200	186,000	198,500	184,200	182,700	215,200	195,000

Source: Own elaboration based on EUROSTAT (2022). *NA* not available

Table 9 Phosphate fertilizer consumption by EU member state (tons)

	2013	2014	2015	2016	2017	2018	2019	2020	2021
Belgium	6509	6919	6254	5635	5485	5670	5835	5651	5657
Bulgaria	11,656	28,144	27,573	36,050	29,562	33,280	33,501	34,441	31,836
Czechia	18,201	20,070	21,221	20,716	24,159	22,459	25,337	20,716	15,783
Denmark	12,300	14,000	13,800	13,800	20,800	14,810	14,767	16,288	15,062
Germany	124,073	123,999	131,503	125,643	100,894	91,047	87,830	108,180	83,910
Estonia	3129	3775	3522	3444	4063	4062	4055	4827	4957
Ireland	36,986	35,584	36,551	37,075	41,893	46,387	42,672	44,259	46,068
Greece	25,896	24,024	21,962	22,515	28,172	25,886	25,977	28,308	30,074
Spain	189,014	174,028	179,784	181,186	190,414	185,983	209,386	212,467	173,028
France	202,345	214,402	198,553	189,602	182,940	181,418	190,283	186,958	189,189
Croatia	12,518	16,875	12,899	5781	14,891	15,564	15,030	15,373	15,025
Italy	98,079	99,748	99,007	99,648	100,273	99,376	98,160	100,037	100,310
Cyprus	1945	1728	2122	2349	2173	2267	2267	NA	NA
Latvia	10,670	10,204	10,633	11,111	11,300	11,481	11,712	13,640	14,309
Lithuania	18,165	19,419	19,799	22,217	23,473	22,425	22,960	23,801	24,110
Luxembourg	515	515	516	491	510	396	371	389	415
Hungary	35,708	35,821	36,020	41,337	51,615	51,152	49,966	48,956	50,531
Malta	31	28	38	56	55	40	40	NA	NA
Netherlands	3984	5869	3736	6011	5416	5271	4957	5702	5752
Austria	14,234	14,110	12,495	15,451	11,789	13,110	13,238	12,208	9757
Poland	163,329	148,914	132,544	142,281	149,950	147,865	149,961	156,627	NA
Portugal	17,289	18,456	20,159	19,037	17,934	16,980	16,645	16,755	18,734
Romania	49,697	51,772	57,921	55,097	63,253	82,264	87,904	81,900	103,728
Slovenia	3886	3991	4120	4000	3988	4116	3538	3668	3183
Slovakia	8959	9918	9426	10,608	10,133	11,209	11,392	11,969	13,036
Finland	11,184	11,845	10,983	9828	12,252	11,033	11,384	11,463	12,761
Sweden	11,800	12,100	12,500	13,100	14,400	14,300	12,800	16,600	17,800

Source: Own elaboration based on EUROSTAT (2022). *NA* not available

Table 10 Consumption of plant protection products by EU member state, expressed in kilograms per year

	2015	2016	2017	2018	2019	2020	2021	2022
Belgium	6,479,244	6,570,443	6,245,917	6,355,292	5,940,604	5,667,462	5,575,943	5,065,853
Bulgaria	N.a	N.a	N.a	N.a	N.a	N.a	4,481,577	4,384,645
Czechia	6,310,835	5,942,725	5,248,376	5,178,116	5,052,758	4,359,751	4,479,836	5,542,760
Denmark	2,583,747	2,586,881	2,706,105	2,646,126	2,660,866	3,167,460	2,974,529	NA
Germany	47,850,638	46,889,399	48,306,978	44,960,615	45,180,977	47,973,603	48,712,164	48,169,069
Estonia	NA	NA	NA	NA	NA	NA	NA	777,074
Ireland	3,503,606	3,112,528	2,942,016	2,651,359	2,971,772	2,909,819	3,090,944	2,715,824
Greece	4,179,705	4,706,652	4,523,331	4,860,470	4,867,463	4,901,423	4,711,803	3,988,154
Spain	NA	76,940,641	71,987,324	73,121,026	75,190,440	75,774,801	76,173,556	NA
France	66,629,333	72,076,247	70,458,871	83,983,125	54,303,708	64,789,218	69,697,520	67,874,942
Croatia	2,310,932	1,861,197	1,577,407	1,697,664	1,563,806	1,655,091	1,469,654	1,507,667
Italy	63,084,973	59,963,868	56,451,019	54,038,535	48,405,281	56,372,591	50,177,305	NA
Cyprus	NA	NA	NA	1,183,607	1,230,773	1,190,578	1,104,019	1,009,795
Latvia	1,510,124	1,724,623	1,478,432	1,586,987	1,650,649	1,900,410	1,984,426	1,822,854
Lithuania	2,788,522	3,385,131	3,056,462	2,081,622	2,351,995	2,587,217	2,612,899	3,614,062
Luxembourg	NA	203,383	189,759	160,308	143,492	147,015	129,696	117,067
Hungary	9,489,182	9,763,859	9,754,223	8,535,054	7,814,977	8,679,090	8,878,589	8,578,155
Malta	NA	NA	NA	NA	NA	NA	108,713	101,379
Netherlands	9,957,697	10,764,575	10,549,857	9,366,131	9,199,841	9,866,878	9,346,731	8,732,216
Austria	3,777,617	4,361,487	4,624,882	5,279,544	4,954,458	5,566,034	5,790,560	6,068,732
Poland	NA	NA	25,073,556	NA	24,253,221	24,616,368	26,949,220	22,299,946
Portugal	10,003,041	9,772,274	8,103,158	8,057,311	9,865,810	9,706,261	9,583,196	9,039,565
Romania	11,483,898	10,812,730	11,548,214	11,107,634	9,046,656	8,700,093	8,369,177	7,957,018
Slovenia	1,046,605	1,155,955	1,087,022	1,171,323	973,221	963,765	907,187	NA
Slovakia	2,243,614	2,092,560	NA	2,490,217	NA	NA	2,275,234	1,979,794
Finland	4,009,553	4,591,602	4,365,909	4,901,621	4,034,153	4,916,941	4,099,762	3,327,273
Sweden	2,321,276	NA	NA	NA	1,800,854	2,047,474	2,137,899	2,338,356

Source: Own elaboration based on EUROSTAT (2024). *NA* not available

and France exhibit relatively high and variable consumption levels through the years, possibly indicative of intensive agricultural methodologies and shifts in pesticide regulatory frameworks. Conversely, Bulgaria and Malta report comparatively lower pesticide use, which may reflect differing agricultural scales or practices.

The variability in data availability across countries and over time suggests disparities in data collection methodologies and reporting standards, as well as the degree of emphasis placed on pesticide monitoring in each nation.

Table 11 details the harmonized risk indicator for pesticides across EU member states, annually. This indicator integrates data on the toxicity of utilized pesticides and their potential exposure risk to humans and the environment. The values of the indicator are presented as an index, with a set base value of 100 corresponding to the reference period of 2011–2013. Notably, Estonia and Bulgaria consistently register

Table 11 Harmonized risk indicator for pesticides by member state. Index value (base period 2011–2013 = 100)

	2015	2016	2017	2018	2019	2020	2021
EU-27	99	94	86	86	79	68	62
Belgium	105	100	84	85	66	48	47
Bulgaria	48	106	82	142	152	84	67
Czechia	91	82	68	69	62	40	37
Denmark	53	53	56	57	57	58	56
Germany	103	99	99	86	84	73	73
Estonia	141	175	160	137	157	144	140
Ireland	114	102	90	83	124	50	52
Greece	58	60	60	63	59	50	43
Spain	105	88	70	74	85	77	65
France	103	103	92	106	61	67	68
Croatia	86	72	62	65	59	54	43
Italy	101	97	95	89	81	66	56
Cyprus	91	91	90	89	90	79	71
Latvia	117	131	116	123	136	120	123
Lithuania	114	138	129	95	105	90	91
Luxembourg	104	77	69	62	52	45	36
Hungary	117	115	111	97	81	76	70
Malta	116	78	87	78	68	78	69
Netherlands	82	88	82	75	68	59	55
Austria	99	111	116	125	118	116	124
Poland	107	105	100	91	106	64	67
Portugal	88	101	79	67	71	60	60
Romania	57	53	60	57	46	36	32
Slovenia	100	107	98	103	75	72	68
Slovakia	92	90	92	98	91	71	58
Finland	121	133	128	140	116	137	115
Sweden	82	67	69	59	56	58	61

Source: Own elaboration based on EUROSTAT (2022)

high indicator values, signaling a relatively elevated risk in comparison to other member states. Conversely, countries like Romania and Luxembourg exhibit lower indicator values, suggesting a diminished relative risk.

4.3.4 Energy Consumption in Agriculture

Table 12 illustrates the analysis of energy consumption within the agricultural sector of the European Union (EU), showcasing significant diversity among member states in terms of energy efficiency and agricultural practices. A review of the data over time uncovers several noteworthy trends and disparities.

Notably, there is considerable variation in energy consumption per hectare across EU countries during the study period. The Netherlands, Belgium, and Malta, for instance, report exceptionally high values, with the Netherlands' energy consumption surpassing 2000 kilograms of oil equivalent per hectare in certain years. Such elevated energy demands can be attributed to the intensification of agriculture in these nations, marked by the use of advanced technologies like greenhouses and highly mechanized irrigation systems, which necessitate significant energy for operation.

Table 12 Energy consumption in agriculture and forestry per unit of area, expressed in kilograms of oil equivalent per hectare

	2012	2013	2014	2015	2016	2017	2018	2019
EU-27	153	156	152	150	154	155	170	
Germany	79	75	91	87	90	85	196	213
Austria	183	191	194	196	203	208	200	197
Belgium	571	612	531	547	579	599	589	659
Bulgaria	39	39	38	37	37	35	37	37
Czech Republic	160	173	175	174	184	182	176	180
Cyprus	318	353	331	308	371	358	320	344
Croatia	156	157	171	137	138	141	147	147
Denmark	252	252	236	241	240	229	226	222
Spain	114	117	113	95	101	101	102	107
Estonia	115	113	134	133	129	134	126	114
Finland	333	335	324	313	323	314	303	301
France	138	146	144	145	140	138	141	140
Greece	55	56	47	49	51	57	50	54
Hungary	75	98	111	108	121	114	121	127
Ireland	55	50	46	45	46	47	49	49
Italy	209	203	210	206	207	217	207	
Latvia	77	79	78	82	87	96	93	103
Lithuania	38	36	35	33	35	37	37	37
Luxemburg	190	201	193	191	221	188	177	174
Malta	346	251	259	171	259	398	420	447
Netherlands	1.999	1998	1844	1957	2031	2042	2159	2.185

Source: Own elaboration based on EUROSTAT (2022)

Conversely, countries such as Germany and France are trending toward more moderate levels of energy consumption relative to other EU member states. Germany, in particular, has seen a notable increase in energy consumption over the years, possibly reflecting an expanded use of energy-intensive technologies or shifts in agricultural production methods.

Spain, contrastingly, demonstrates comparatively lower energy consumption levels, maintaining around 100–120 kilograms of oil equivalent per hectare in most years, despite hosting one of the largest greenhouse systems in the world. This efficiency is attributed to the advantageous climate, which reduces the necessity for climate correction equipment (Valera-Martínez et al., 2017). Nonetheless, Spain has experienced consumption fluctuations throughout the study period, likely due to variations in agricultural practices, energy policies, or climatic conditions.

4.3.5 Organic Crop Area

Table 13 outlines the percentages of land allocated to organic farming within various EU member states from 2015 to 2021. This information is pivotal for gauging the development and prominence of organic farming across Europe, a method of agriculture that fosters environmental sustainability, enhances biodiversity, and minimizes reliance on synthetic chemicals. The growth of organic farming is a focal point for the EU's agricultural policy objectives by 2030 (European Commission, 2020a).

A notable variability exists among member states in the extent of land committed to organic agriculture. Austria emerges as a consistent frontrunner, with more than 20% of its total agricultural area dedicated to organic farming in most surveyed years, underscoring its dedication to sustainable agricultural practices and organic production.

Similarly, other nations demonstrate a significant dedication to organic farming. Estonia, for instance, has shown a continuous rise in the proportion of land devoted to organic agriculture, indicating an increasing commitment to sustainable farming practices within the country. Likewise, Sweden, Finland, and Denmark consistently report high percentages, illustrating a regional shift toward more eco-friendly farming methods in the Scandinavian countries.

Conversely, some countries exhibit lower proportions of land devoted to organic agriculture. Malta, for example, registers the smallest percentages, possibly attributed to its geographic and size constraints, coupled with unique agricultural challenges associated with small island states.

Table 13 Proportion of organic crops in the total cultivated area of each EU member state, expressed as a percentage

	2015	2016	2017	2018	2019	2020	2021
EU-27	6.56	7.09	7.47	7.99	8.47	9.1	NA
Belgium	5.17	5.8	6.28	6.56	6.85	7.25	7.48
Bulgaria	2.37	3.2	2.72	2.56	2.34	2.3	1.71
Czechia	13.68	14	14.09	14.76	15.19	15.33	15.55
Denmark	6.33	7.81	8.6	9.75	11.09	11.45	11.58
Germany	6.34	6.82	6.82	7.34	7.75	9.59	9.65
Estonia	15.68	18.02	20.01	20.98	22.33	22.41	22.97
Ireland	1.65	1.72	1.66	1.65	1.63	1.66	2
Greece	7.69	6.51	7.96	9.32	10.26	10.15	NA
Spain	8.24	8.48	8.73	9.28	9.66	9.98	10.79
France	4.54	5.28	5.99	7.01	7.72	8.77	9.67
Croatia	4.94	6.05	6.46	6.94	7.19	7.21	8.26
Italy	11.79	13.99	14.67	15.17	15.16	15.96	16.83
Cyprus	3.72	4.94	4.61	4.55	4.98	4.63	6.43
Latvia	12.29	13.42	13.92	14.47	14.79	14.79	15.34
Lithuania	7.11	7.5	7.98	8.13	8.14	8	8.91
Luxembourg	3.21	3.47	4.15	4.39	4.42	4.63	5.19
Hungary	2.43	3.48	3.73	3.92	5.71	6.03	5.81
Malta	0.25	0.21	0.35	0.41	0.47	0.62	0.61
Netherlands	2.67	3.03	3.31	3.5	3.75	3.95	4.22
Austria	20.3	21.25	23.37	24.08	25.33	25.69	N.a
Poland	4.03	3.72	3.41	3.33	3.49	3.45	3.78
Portugal	6.53	6.74	6.84	5.68	7.39	8.05	19.31
Romania	1.77	1.67	1.93	2.43	2.86	3.59	4.42
Slovenia	8.85	9.12	9.6	10.01	10.35	10.29	10.81
Slovakia	9.47	9.75	9.9	9.85	10.31	11.67	13.45
Finland	9.91	10.47	11.41	13.09	13.48	13.93	14.45
Sweden	17.14	18.3	19.16	20.29	20.43	20.31	20.2

Source: Own elaboration based on EUROSTAT (2022). *NA* not available

5 Conclusions

This book chapter concludes that the European Union (EU) member states' agri-food sector exhibits a moderate sustainability level. Among the rankings, France emerges as the leader, closely followed by Austria and Italy. Several countries, including Estonia, Germany, Belgium, Finland, Denmark, Spain, Lithuania, the Czech Republic, Sweden, Greece, the Netherlands, and Slovakia, outperform the EU average in terms of agri-food sector sustainability. Conversely, Malta, Cyprus, Bulgaria, and Hungary register the lowest sustainability levels within this sector, with no EU member state achieving the theoretical maximum of the model.

The analysis reveals diverse sustainability statuses across the economic, social, and environmental components within the EU's agri-food sector. Literature review identifies a sustainability classification pattern across six groups, indicating varied strengths among member states in specific indicators. This diversity underscores the necessity for tailored agricultural policies to enhance agri-food sustainability levels in individual countries. Economically, policies aimed at boosting agri-food exports could enhance incomes and thereby improve the sector's social dimensions. Furthermore, enforcing measures to reduce the environmental footprint of food and beverage production and processing is crucial, without compromising productivity. The promotion of circular economy principles can sustain productivity and economic benefits of agri-food systems while enhancing their environmental sustainability.

In essence, this chapter offers a comprehensive and nuanced overview of the EU agri-food sector's sustainability, highlighting strengths, pinpointing challenges, and identifying opportunities for sustainability enhancement. These insights are vital for shaping future policies and strategies to foster a more sustainable and equitable development within the European agri-food sector. The agri-food sector of the European Union (EU) member states exhibited a moderate level of sustainability, according to the results gathered in this book chapter. In the ranking, France ranked first, followed by Austria and Italy. Countries like Estonia, Germany, Belgium, Finland, Denmark, Spain, Lithuania, Czech Republic, Sweden, Greece, Netherlands, and Slovakia surpassed the average of countries in the European Union in terms of sustainability in the agri-food sector. On the other hand, Malta, Cyprus, Bulgaria, and Hungary showed the lowest level of overall sustainability in this sector. These results indicate that none of the EU member states reached the theoretical maximum of the model.

The status of each component of sustainability in the agri-food sector of the EU countries (i.e., economic, social, and environmental) differs. According to the literature, a pattern of sustainability classification into six different groups has been observed, as each member state excels in different simple indicators. This indicates the need to implement agricultural policy with diverse approaches to improve the level of agri-food sustainability in each member state. In general, economic policies that enhance agri-food exports should be expanded, which will improve income and, consequently, the social aspect of the agri-food sector. Additionally, policies promoting the reduction of the environmental footprint of food and beverage production and transformation need to be enforced while maintaining productivity. Promoting techniques based on the circular economy can help maintain productivity and, therefore, the economic benefit of agri-food systems while expanding their environmental sustainability.

In summary, this chapter provides a broad and detailed overview of sustainability in the EU agri-food sector, identifying areas of strength, challenges, and opportunities for improving sustainability in all aspects. These findings are critical to inform future policies and strategies aimed at promoting more sustainable and equitable development in the European agri-food sector.

References

Abbate, S., Centobelli, P., Cerchione, R., Giardino, G., & Passaro, R. (2023). Coming out the egg: Assessing the benefits of circular economy strategies in agri-food industry. *Journal of Cleaner Production, 385*, 135665. https://doi.org/10.1016/j.jclepro.2022.135665

Ayompe, L. M., Schaafsma, M., & Egoh, B. N. (2021). Towards sustainable palm oil production: The positive and negative impacts on ecosystem services and human wellbeing. *Journal of Cleaner Production, 278*, 123914. https://doi.org/10.1016/j.jclepro.2020.123914

Bonisoli, L., Galdeano-Gómez, E., Piedra-Muñoz, L., & Pérez-Mesa, J. C. (2019). Benchmarking agri-food sustainability certifications: Evidences from applying SAFA in the Ecuadorian banana agri-system. *Journal of Cleaner Production, 236*, 117579. https://doi.org/10.1016/j.jclepro.2019.07.054

Camacho-Ferre, F. (2004). Técnicas de producción de cultivos protegidos (Tomo I). *Caja Rural Intermediterránea, Cajamar, Almería, España*, 1–373.

Castillo-Díaz, F. J., Belmonte-Ureña, L. J., Camacho-Ferre, F., & Tello-Marquina, J. C. (2022). Biodisinfection as a profitable fertilization method for horticultural crops in the framework of the circular economy. *Agronomy, 12*, 521. https://doi.org/10.3390/agronomy12020521

Castillo-Díaz, F. J., Belmonte-Ureña, L. J., López-Serrano, M. J., & Camacho-Ferre, F. (2023). Assessment of the sustainability of the European agri-food sector in the context of the circular economy. *Sustainable Production and Consumption, 40*, 398–411. https://doi.org/10.1016/j.spc.2023.07.010

Christiaensen, L., Demery, L., & Kuhl, J. (2011). The (evolving) role of agriculture in poverty reduction – An empirical perspective. *Journal of Development Economics, 96*(2), 239–254. https://doi.org/10.1016/j.jdeveco.2010.10.006

Cifuentes-Faura, J. (2022). European Union policies and their role in combating climate change over the years. *Air Quality, Atmosphere and Health, 15*(8), 1333–1340. https://doi.org/10.1007/s11869-022-01156-5

Correia, J., João, P., Lopes, I., Duarte, A. C., & Rocha-Santos, T. (2020). Science of the total environment environmental exposure to microplastics: An overview on possible human health effects. *Science of the Total Environment, 702*, 134455. https://doi.org/10.1016/j.scitotenv.2019.134455

Davies, C., Chen, W. Y., Sanesi, G., & Lafortezza, R. (2021). The European Union roadmap for implementing nature-based solutions: A review. *Environmental Science and Policy, 121*(March), 49–67. https://doi.org/10.1016/j.envsci.2021.03.018

De Carvalho, M. I., Relvas, S., & Barbosa-Póvoa, A. P. (2022). A roadmap for sustainability performance assessment in the context of agri-food supply chain. *Sustainable Production and Consumption, 34*, 565–585. https://doi.org/10.1016/j.spc.2022.10.001

Dixon, J., Gulliver, A., & Gibbon, D. (2001). Sistemas de Producción Agropecuaria y Pobreza. In *FAO y el Banco Mundial. Roma*. http://www.fao.org/3/a-ac349s.pdf

Duque-Acevedo, M., Belmonte-Ureña, L. J., Plaza-Úbeda, J. A., & Camacho-Ferre, F. (2020). The management of agricultural waste biomass in the framework of circular economy and bioeconomy: An opportunity for greenhouse agriculture in Southeast Spain. *Agronomy, 10*(4). https://doi.org/10.3390/agronomy10040489

Egea, F. J., Torrente, R. G., & Aguilar, A. (2017). An efficient agro-industrial complex in Almería (Spain): Towards an integrated and sustainable bioeconomy model an efficient agro-industrial complex in Almería (Spain): Towards an integrated and sustainable bioeconomy model. *New Biotechnology, 40*(January 2018), 103–112. https://doi.org/10.1016/j.nbt.2017.06.009

Elsevier. (2023). *Scopus*. https://www.elsevier.com/es-es/solutions/scopus

European Commission. (2019). The European green deal. *Office of the European Union. Brussels, Belgium*, 1–28.

European Commission. (2020a). A farm to fork strategy for a fair, healthy and environmentally-friendly food system. *Office of the European Union. Brussels, Belgium*, 1–23.

European Commission. (2020b). A new circular economy action plan. In for a cleaner and more competitive Europe. *Office of the European Union. Brussels, Belgium*, 23. https://eur-lex.europa.eu/resource.html?uri=cellar:9903b325-6388-11ea-b735-01aa75ed71a1.0018.02/DOC_1&format=PDF
European Commission. (2020c). EU biodiversity strategy for 2030. Bringing nature back into our lives. *Office of the European Union. Brussels, Belgium*, 1–27.
European Parliament. (2021). REGULATION (EU) 2021/1119 OF THE EUROPEAN PARLIAMENT AND OF THE COUNCIL of 30 June 2021 establishing the framework for achieving climate neutrality and amending Regulations (EC) No 401/2009 and (EU) 2018/1999 ('European Climate Law'). *Official Journal of the European Union, L 243*, 1–17.
Eurostat. (2022). *Eurostat*. https://ec.europa.eu/eurostat
FAO. (2017a). El estado mundial de la agricultura y la alimentación. *Organización de Las Naciones Unidas Para La Agricultura y La Alimentación, Roma, Italia*, 1–2017.
FAO. (2017b). *Towards Zero Hunger 1945–2030*. Food and Agriculture Organization of the United Nations.
FAO. (2019a). *Background notes on sustainable, productive and resilient agro-food systems: Value chains, human capital, and the 2030 agenda* (Issue July). Food and Agriculture Organization of the United Nations.
FAO. (2019b). *The international for the sustainable use Code of Conduct and management of fertilizers* (Food and A).
FECYT. (2022). *Base de datos Scopus*. https://www.recursoscientificos.fecyt.es/licencias/productos-contratados/scopus#:~:text=Scopusesunabasede,yvisualizacióndelainvestigación
Fernández, E., Ramón-Díez, J., Aspachs, O., Jódar, S., & Montoriol, J. (2023). *INFORME SECTORIAL Agroalimentario 2023*. CaixaBank Research.
Gómez-Tenorio, M. Á., Magdaleno-González, J., & Tello-Marquina, J. C. (2021). *Evaluación e implementación de técnicas regenerativas para la mejora de la fertilidad en el cultivo del almendro en las provincias de Almería y Granada*. Editorial Tecnoagrícola de España.
Green, A., Nemecek, T., Chaudhary, A., & Mathys, A. (2020). Assessing nutritional, health, and environmental sustainability dimensions of agri-food production. *Global Food Security, 26*(January), 100406. https://doi.org/10.1016/j.gfs.2020.100406
Honoré, M. N., Belmonte-Ureña, L. J., Navarro-Velasco, A., & Camacho-Ferre, F. (2019). Profit analysis of papaya crops under greenhouses as an alternative to traditional intensive horticulture in Southeast Spain. *International Journal of Environmental Research and Public Health, 16*(16). https://doi.org/10.3390/ijerph16162908
Kirchherr, J., Reike, D., & Hekkert, M. (2017). Conceptualizing the circular economy: An analysis of 114 definitions. *Resources, Conservation and Recycling, 127*(April), 221–232. https://doi.org/10.1016/j.resconrec.2017.09.005
López-Serrano, M. J., Velasco-Muñoz, J. F., Aznar-Sánchez, J. A., & Román-Sánchez, I. M. (2020). Sustainable use of wastewater in agriculture: A bibliometric analysis of worldwide research. *Sustainability (Switzerland), 12*(21), 1–20. https://doi.org/10.3390/su12218948
MAPA, & Cajamar. (2022). *Diagnóstico inicial del estado de la transformación digital del sector agroalimentario español*. Ministerio de Agricultura Pesca y Alimentacion y Cajamar.
MAPA, & Cajamar. (2023). *Análisis del estado actual de la digitalización del sector agroalimentario español*. Ministerio de Agricultura Pesca y Alimentacion y Cajamar.
Martínez-Francés, M. A., Lacasa-Plasencia, A., & Tello-Marquina, J. C. (2009). *Ecología de la microbiota fúngica de los suelos de los invernaderos de pimiento y su interés agronómico*. Ministerio de Agricultura Pesca y Alimentacion.
Maudos, J., & Salamanca, J. (2022). *Observatorio sobre el sector agroalimentario español en el contexto europeo Informe 2021*. Cajamar Caja Rural e IVIE.
Maudos, J., & Salamanca, J. (2023). *Observatorio sobre el sector agroalimentario español en el contexto europeo Informe 2022*. Cajamar Caja Rural e IVIE.
Mundial, B. (2021). *Valor agregado de la agricultura*. Banco Mundial. https://datos.bancomundial.org/indicator/NV.AGR.TOTL.ZS

OCDE-FAO. (2019). *OCDE-FAO Perspectivas Agrícolas 2019–2028*. OECD Publishing. https://doi.org/10.1787/7b2e8ba3-es.

Oehlmann, J., Schulte-Oehlmann, U., Kloas, W., Jagnytsch, O., Lutz, I., Kusk, K. O., Wollenberger, L., Santos, E. M., Paull, G. C., VanLook, K. J. W., & Tyler, C. R. (2009). A critical analysis of the biological impacts of plasticizers on wildlife. *Philosophical Transactions of the Royal Society B: Biological Sciences, 364*(1526), 2047–2062. https://doi.org/10.1098/rstb.2008.0242

Orou, R., Kirschke, S., & Günther, E. (2023). Integrating the social perspective into the sustainability assessment of agri-food systems: A review of indicators. *Sustainable Production and Consumption, 39*(April), 175–190.

Pankhurst, C. E., & Lynch, J. M. (1995). The role of soil microbiology in sustainable intensive agriculture. *Advances in Plant Pathology, 11*(C), 229–247. https://doi.org/10.1016/S0736-4539(06)80014-0

Paris, B., Vandorou, F., Balafoutis, A. T., Vaiopoulos, K., Kyriakarakos, G., Manolakos, D., & Papadakis, G. (2022). Energy use in open-field agriculture in the EU: A critical review recommending energy efficiency measures and renewable energy sources adoption. *Renewable and Sustainable Energy Reviews, 158*(August 2021), 112098. https://doi.org/10.1016/j.rser.2022.112098

Pedraza, A. C., Díaz, A. R., & Soto, I. E. (2015). Cambios paisajísticos y efectos medioambientales debidos a la agricultura intensiva en la Comarca de Campo de Cartagena-Mar Menor (Murcia). *Estudios Geograficos, 76*(279), 473–498. https://doi.org/10.3989/estgeogr.201517

Ruiz-Almeida, A., & Rivera-Ferre, M. G. (2019). Internationally-based indicators to measure agri-food systems sustainability using food sovereignty as a conceptual framework. *Food Security, 11*, 1321–1337. https://doi.org/10.1007/s12571-019-00964-5

Silvestri, C., Silvestri, L., Piccarozzi, M., & Ruggieri, A. (2022a). *Toward a framework for selecting indicators of measuring sustainability and circular economy in the agri-food sector: A systematic literature review*. Springer.

Silvestri, L., Silvestri, C., Forcina, A., & De Luca, C. (2022b). A review of energy-based indicators for assessing sustainability and circular economy in the agri-food production. *Procedia Computer Science, 200*(2019), 1756–1765. https://doi.org/10.1016/j.procs.2022.01.376

Streimikis, J., & Baležentis, T. (2020). Agricultural sustainability assessment framework integrating sustainable development goals and interlinked priorities of environmental, climate and agriculture policies. *Sustainable Development, 28*(6), 1702–1712. https://doi.org/10.1002/sd.2118

The World Bank. (2008). *World development report: Agriculture Development* (Issue c).

UN. (2015). Draft outcome document of the United Nations summit for the adoption of the post-2015 development agenda. In *The World Bank Group A to Z 2016* (Vol. 13689). 10.1596/978-1-4648-0484-7_post_2015_development_agenda.

Valera-Martínez, D. L., Belmonte-Ureña, L. J., Molina Aiz, F. D., & Camacho-Ferre, F. (2017). The greenhouses of Almería, Spain: Technological analysis and profitability. *Acta Horticulturae, 1170*, 2019–2226. https://doi.org/10.17660/ActaHortic.2017.1170.25

Velten, S., Leventon, J., Jager, N., & Newig, J. (2015). What is sustainable agriculture? A systematic review. *Sustainability, 7*, 7833–7865. https://doi.org/10.3390/su7067833

Zapata Sierra, A., Zapata Castillo, L., & Manzano Agugliaro, F. (2022). Water resources availability in southern Europe at the basin scale in response to climate change scenarios. *Environmental Sciences Europe, 34*, 75. https://doi.org/10.1186/s12302-022-00649-5

Index

S. S. Muthu (ed.), *Assessment of Social Sustainability Management in Various Sectors*, Environmental Footprints and Eco-design of Products and Processes, https://doi.org/10.1007/978-3-031-81029-9

The manufacturer's authorised representative in the EU is Springer Nature Customer Service Centre GmbH, Europaplatz 3, 69115 Heidelberg, Germany. If you have any concerns regarding our products, please contact ProductSafety@springernature.com

Printed and bound by CPI Group (UK) Ltd, Croydon, CR0 4YY
01/10/2026
02225948-0002